MINISTRY OF
AGRICULTURE, FISHERIES AND FOOD

AF845999

Soils and Manures for Fruit

Bulletin 107

LONDON
HER MAJESTY'S STATIONERY OFFICE
1975

© *Crown copyright 1975*

First published 1937
Fourth edition 1975

ISBN 0 11 241407 9

Contents

Foreword

THE importance of choosing suitable soils for fruit cannot be over emphasized in view of the large capital investment required to establish a fruit enterprise and the need in today's economic climate to produce large yields of good quality fruit. This bulletin provides a useful guide to soils, not only to intending fruit growers but also to those already established. Guidance is also given on soil management including the use of fertilizers and manures, subjects requiring careful and continuous consideration if growers are to avoid major problems.

This is the fourth edition of the bulletin, which was first published in 1937, and incorporates recent developments on soils and manures for fruit. It was prepared by the following group appointed by the Soil Scientists' Publication Committee:

Dr. R. C. Little, A.D.A.S. Regional Soil Scientist, West Midland Region.

Mr. R. F. Clements, A.D.A.S. Regional Fruit Adviser, South Eastern Region.

Mr. D. J. Eagle, A.D.A.S. Soil Scientist, Eastern Region.

Ministry of Agriculture, Fisheries and Food

March, 1974

Soils for Fruit

In the past, the soils chosen for the establishment of orchards in this country were usually those in a high state of fertility or brought to high fertility a few years prior to planting. In cultivated orchards great care was taken to preserve and improve soil condition by careful cultural operations. Old hop gardens in Kent where the fertility of the soil had been raised to a high level, were often chosen for the establishment of fruit plantations. In the West Midlands, fruit growing spread to the Avon Valley from the Abbey gardens at Evesham, and Pershore, again on highly fertile soils, and in the area west of Worcester and in the Teme Valley on the fertile marl soils some of which had previously been used for hops. In East Anglia, the early plantings in the Wisbech area were on the deep silt soils, noted for their fertility.

The growing of fruit also took place in areas where it fitted well into the farming system. The apple orchards of Gloucestershire, Herefordshire, Somerset and Devon are examples. These were established in grass fields, again often on naturally fertile soils near farmsteads. The grass of the orchards was grazed throughout the year by cattle and sheep, thus allowing livestock and fruit enterprises to run side by side. In a few of the early developments, sites were chosen not so much for suitability of their soils but because they were near appropriate markets or where early crops could be produced. So it was that in the Southampton area many rather unsuitable soils were used for growing early crops of fruit, particularly early strawberries.

In the expansion of these older districts and the rapid development of new areas of fruit growing that has taken place, particularly since 1920, there has been a tendency to plant fruit on soils of much lower fertility than those previously used, including poor sands and gravel and relatively shallow or heavy soils with poor drainage. Indeed, at one stage the idea was current that the planting of some fruit crops, particularly dessert apples, was a good method of utilizing poor soils. Many of the light gravelly soils of Essex were specially selected as being particularly suitable for Cox's Orange Pippin apples, and today the great variability in growth of trees on some of these gravel soils with inherent soil defects, can still be seen. However, orchards on such soils were often economic due to profit margins being much greater than in recent years.

It is now possible to define the soil requirements of fruit crops with some precision and in the present economic climate of small profit margins it is essential to plant only soils which can give optimum production. Perennial fruit crops require soils which hold ample moisture within rooting depth but which are also free from waterlogging throughout the whole year. However requirements vary to some extent. Raspberries for example require a higher standard of drainage than pears or culinary apples. The requirements of the common fruit crops are described in subsequent sections and the occurrence and correction of soil defects is discussed.

Choice of Site

When choosing an area for fruit, careful consideration must first be given to climate, altitude, topography, aspect, slope and proximity to large bodies of water including the sea.

As most fruit crops blossom during April or May, when frosts are common, areas where cold air can accumulate must be avoided since frost at blossom time can be seriously damaging. Cold air tends to flow downwards and collect in depressions so the incidence of frost, especially in spring, is closely associated with the contour of the land. Low lying areas are frost susceptible when there is no outlet for 'air drainage' to still lower land. The risk of damage from frost can be minimized by choosing a site with good air drainage. Such sites are usually found on upper slopes but it is not sufficient to have an elevated site; there must also be no impedence to air drainage by, for example, woodland or hedges across the slope. Actual height is less important than comparative height above surrounding land. Low lying land is not necessarily frosty if cold air can flow away freely.

Proximity to water tends to reduce the frost risk as the mass of water cools less rapidly than the land and helps to maintain night temperatures above freezing. In the warmer months, however, the day temperatures will be rather lower. Exposure to salt laden winds from the sea may result in damage to foliage and russeting of apples.

Sites exposed to strong winds should be avoided as wind can damage leaves or blossom and autumn gales can blow off the fruit. Wind rocking of trees can damage roots and puddle the soil at the base of the trunk; water seeping down the hole created by the rocking can collect above the puddled soil and further damage the roots by waterlogging. In the east and south-east of England, north-east and east winds in April and May may cause long cool periods which retard growth. Wind velocity tends to increase with altitude but this may be greatly affected by topography and aspect.

In general, both temperature and sunshine decrease with altitude and land above about 400 ft is not suitable. Except in the case of vines, which need a southerly slope, aspect is generally not of great importance provided steep slopes and exposure to gales are avoided.

To be satisfactory for fruit, sites which are suitable climatically must also be free of soil defects such as poor drainage or susceptibility to drought.

The Soil

The demands that fruit crops make from the soil are simply for water and nutrients but a number of factors affect growth of the roots which absorb them. The nutrients can be easily supplied in the form of fertilizers as and when required but to absorb them an efficient root system is essential. For absorption of water a healthy root system is again necessary as well as an adequate supply of moisture which depends on soil texture and depth.

To ensure an efficient root system the soil must contain a supply of air all the year round and allow carbon dioxide, produced by root respiration, to escape. Roots need oxygen and its exclusion results in the accumulation of toxic substances which can be lethal. Consequently soils must be sufficiently well drained to preclude waterlogging and must be free from compacted layers which impede the movement of air and water. The effects of root damage caused by temporary waterlogging may be severe and can often be seen for up to three years subsequently. In severe cases death of tree or bush can result.

The main soil factors which influence root development are (i) depth of soil (ii) soil texture (iii) soil structure and its stability (iv) the presence of physical obstructions (v) soil drainage (vi) soil pH and nutrient status.

Depth of Soil

The volume of soil in which roots can grow freely without obstruction by barriers of impervious soil, bad drainage, or by lack of moisture or air has a major influence on the vigour and development of roots. The depth of soil that is required to maintain satisfactory fruit crops depends to a considerable extent on the type and variety of fruit and its rootstock. The term 'depth of soil' usually refers to the depth of weathered soil in which conditions are suitable for the normal growth of roots. Where there are poor conditions in the subsoil due, for example, to bad drainage or pans, the depth is taken as the soil above the poor conditions. A deep soil can become shallow by the development of bad soil conditions and soils which are shallow due to similar adverse conditions can often be deepened by correction of these defects. Any consideration of depth of soil must also take into account the influence of the underlying material as this may be impervious and restrict roots, or may allow penetration of some roots, or very often may act as a reservoir for moisture. A shallow depth of soil over chalk may provide satisfactory conditions for reasonable growth of crops because of the water-bearing character of the chalk but there will be a risk of iron deficiency. Similarly, some roots can often penetrate into underlying fissured material and obtain supplies of moisture. This is illustrated in the diagram on page 4.

Fruit trees in particular succeed best on deep soils and are more likely to remain usefully productive over a long period of years when they have well-developed roots. Soils formed from glacial drift materials, wind-bown deposits such as the brickearth soils, downwash soils as are found at the bottom of slopes, and alluvium found in river valleys, are often the deepest types. There is usually no resistance to the auger in these soils to considerable depth, indicating an open structure suitable for root development. On the other hand downwash formed from coarse textured materials may be too open and loose and trees may be poorly anchored. Under these conditions rootstocks which develop strong rooting systems should be used. The effective depth of soil may be small in soils formed *in situ* from older geological materials. These soils are difficult to auger due to compaction of the parent material during its long history.

It is fairly common practice to establish orchards on slopes and as erosion is more likely to occur here it is important to assess its effects. Eroded areas are often shallow and may be compacted and badly drained. In contrast, soils at the bottom of slopes are usually deep, due to downwash and often produce very large trees.

A guide to the requirements of fruit for various depths of soil is given in Table 1. Minimum depths of soil are given but it is important to take into consideration the underlying material.

Soil Texture

The texture of the soil depends on its mineral composition in relation to the grades and amounts of sand, silt and clay present. Texture is modified by the presence of stones and gravel, by high amounts of chalk and limestone and by large amounts of organic matter. The texture of the soil has a fundamental influence on many of the soil properties such as drainage, stability of

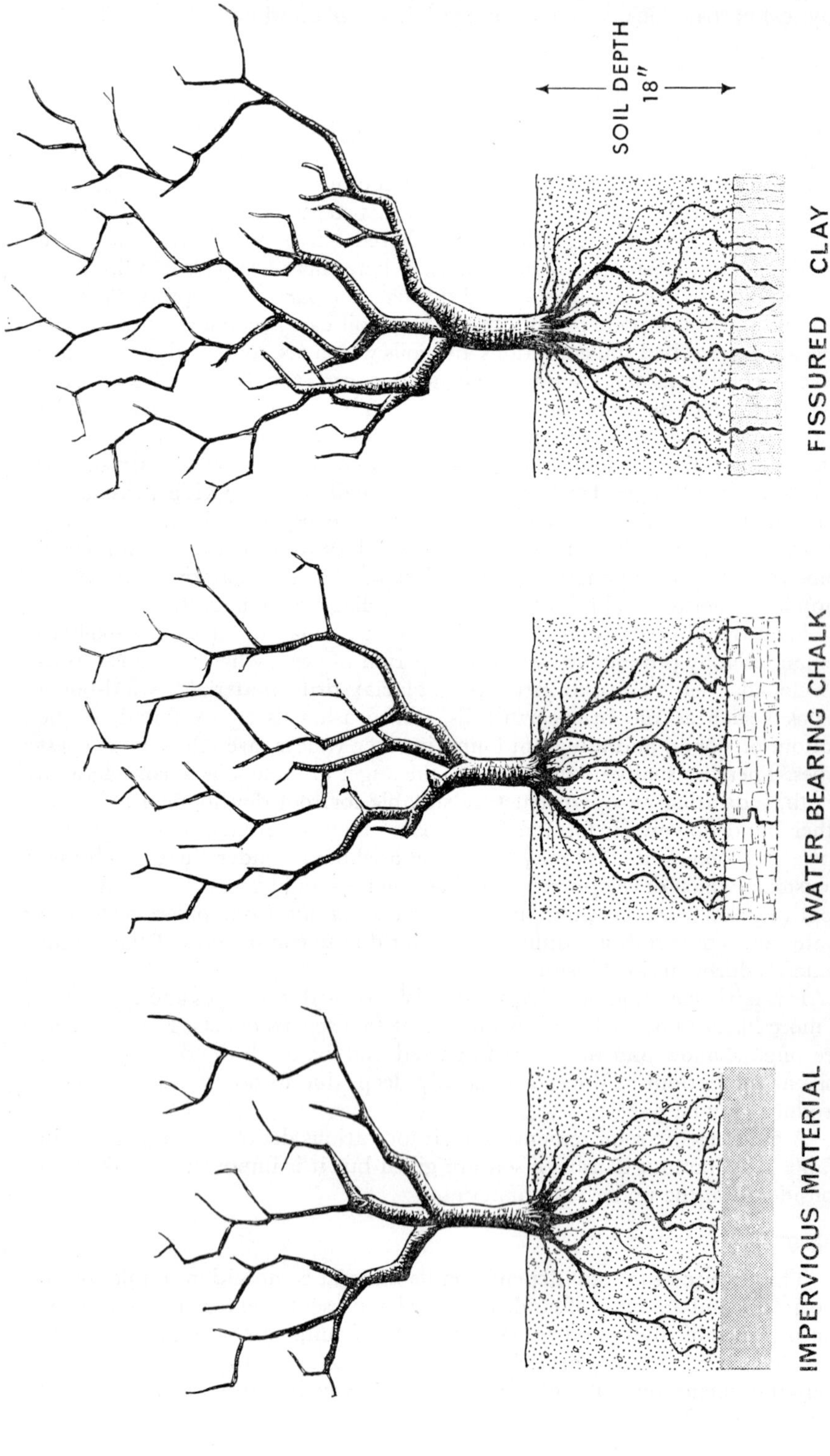

Influence of Underlying Geological Material on Tree Growth

TABLE 1

Type of Fruit and Depth of Soil

Minimum depth of well drained soil, in inches	Description of depth	Fruit and Soil Type
15	Shallow	Unsuitable except perhaps for strawberries. Sandy soils are subject to drought and clayey soils to poor drainage.
18	Moderate	Suitable for most fruits provided there can be some rooting below this depth; raspberries on the sandy loams to loams, gooseberries and red currants on a wider range from sandy loams to clay loams.
24	Deep	Suitable for all soft and top fruits on a wide range of soil textures with the exception of sweet cherries. Dessert apples may be over-vigorous in very deep loam to clay loam soils if a vigorous rootstock is used.
36	Very deep	Sweet cherries require a very deep soil and succeed well on fine sandy loams to loams. (Sour cherries can succeed with less soil depth than sweet cherries).

soil structure, the sizes and amount of pore spaces for air and moisture, and ease of movement of water and roots through soils. The great influence of texture on the amount of water available in the soil for plants at field capacity is illustrated in Table 2. The data refer to soils at field capacity which is the moisture content after the excess has drained away following a period of heavy rain.

TABLE 2

Water available for plants on various soils at field capacity

Soil Texture	Available water per ft depth of soil
	in.
Loamy coarse sand	1·3
Loamy sand	1·9
Loamy fine sand	2·3
Loamy very fine sand . .	2·6
Coarse sandy loam	1·5
Sandy loam	2·1
Fine sandy loam	2·3
Very fine sandy loam . .	2·6
Loam	2·1
Silt loam	2·3
Sandy clay loam	1·8
Clay loam	2·2
Silty clay loam. . . .	2·3

Although the available water capacity becomes less critical when irrigation is available it is generally unwise to plant on very droughty soils because irrigation would be needed very frequently in dry weather. Any interruption in the supply of water could have very serious effects in hot dry weather on soils which can hold sufficient moisture for only a week or so.

Assessing the texture of the Soil

As well as determining the amount of water available per foot of soil the texture of the soil is also important because of its influence on such factors as drainage and stability of the soil structure. Assessing the texture of the soil is of considerable value as it provides important information on the properties of the soil and how it will react under different conditions. The texture of the soil is classified on the basis of the relative amounts of sand, silt and clay present and its assessment can be made in the field by rubbing the moist soil between the thumb and forefinger. Sand can be detected by its grittiness; silt when moist has a smooth silky feel like butter, whereas moist clay takes a polish when rubbed between the fingers and thumb and is plastic. With experience it soon becomes easy to distinguish predominantly sandy, silty or clayey soil.

Loamy Sands

Soils in this group consist mainly of sand but have a small amount of 'body' or loaminess. The soil is predominantly gritty but will bind when a ball of moist soil is made although the ball crumbles and falls apart easily. Loamy sand soils and particularly loamy coarse sand soils are very open in structure, normally low in organic matter and nutrients and are drought prone. Although easily worked and capable of producing early crops they are unsuitable, particularly when shallow, for most classes of fruit. On the deep types where there is also adequate water, soft fruits can be successful. Irrigation or mulching is usually of great value on these soils. Loamy fine and very fine sands are much more moisture retentive and when deep are quite suitable for many fruit crops.

Loams

Loams are soils which bind together without difficulty when moist and the ball formed does not readily fall apart. Sandy loams can be distinguished from loams by their grittiness, silty loams by their smoothness, and clay loams by their ability to take a polish and mould.

Sandy loams

There is a range of sandy loams, depending on the size of the sand grain:

Coarse sandy loams
Sandy loams (medium-sized grain)
Fine sandy loams
Very fine sandy loams

Coarse sandy loams are drought prone but when deep are suitable for some soft fruits. The medium sandy loams are also light and open in structure but are less droughty. They are early soils and are often chosen for strawberries. Dessert apples can be grown successfully on the deeper types where there is adequate moisture. Fine and very fine sandy loams when deep and well drained are very suitable for soft fruit, dessert apples and for cherries. The brick-earth soils, which are noted for their high fertility when well drained are usually very fine sandy loams or silty loams in texture. Fine sandy soils are very liable to pack tightly, particularly when their organic matter content is low, causing poor aeration. Tightly packed soil can lead to bad drainage conditions in and above the consolidated area. Collapse of surface structure in clean cultivated orchards is possible under the action of heavy rain showers and from the movement of equipment over the soil. Considerable care in cultural operations and attention to drainage is therefore necessary with fine sandy soils. However, when weed control is achieved by the use of herbicides alone these soils can build up a satisfactory stable structure within the row.

Silty Loams

The predominance of the fine sandy and silty fractions make these soils close-structured. Unless organic matter is maintained at a moderately high level the soil crumbs break down under clean cultivation and the sand and silt separate. The silt may flow under wet conditions and set when dry into hard layers or crusts. As with fine sandy loams and very fine sandy loams, drainage can be defective in these finely textured soils and stability of structure is generally poor. Nevertheless when conditions are good and maintained with care, these silty loams are very productive and suited for many classes of fruit, as can be seen in the deep, productive soils of the Wisbech fruit area in East Anglia. Where more clay is present (silt loam) culinary apples, pears and plums can be grown satisfactorily provided there is sufficient depth of well drained soil. Weed control by herbicides can result in a satisfactory structure within the row on silty soils as for the fine sandy soils discussed above.

Loams

The loam is intermediate in texture between the sandy loam and clay loam. Loam soils, with proper management, are suitable for all types of fruit and when deep and well drained are very good fruit soils.

Clay Loams

A ball of moist clay loam rubbed between the thumb and forefinger will take a strong polish; the ball binds and moulds and there is some resistance to deformation. Sandy clay loams can be distinguished by some grittiness and silty clay loams by smoothness and are less stable than 'straight' clay loams. Soils of this group are very retentive of moisture and are sticky when wet. When they dry out they often form hard clods which can only be broken down with difficulty. Fruit can be grown successfully on clay loams provided drainage and depth of soil are adequate and the right types of fruit are chosen. Drainage is generally better on slopes than on level sites. Pears and plums are often more successful on good clay loams than on light soils. Culinary apples can also be grown with success on these heavy soils.

Clays

A moist ball of clay is plastic, takes a very strong polish when rubbed and requires considerable pressure to mould between the fingers. Clays are very sticky and difficult soils when wet and dry into hard clods. Very severe cracking of the soil can occur during dry periods. These soils cannot be recommended for fruit. Clays are seldom found as top soils and if present near the surface may indicate very eroded sites which would be very unsuitable for fruit.

A guide to the mineral composition of some examples of soils in the main textural classes previously described is given in Table 3.

TABLE 3

Texture	Percentage Composition			
	Coarse Sand 2–0·2 mm	Fine Sand 0·2–0·02 mm	Silt 0·02–0·002 mm	Clay 0·002 mm
Loamy coarse sand	60	30	5	5
Loamy sand	45	45	5	5
Sandy loam	45	35	10	10
Very fine sandy loam	0	65	15	20
Silty loam	0	50	30	20
Loam	35	25	15	25
Sandy clay loam	20	45	10	25
Clay loam	10	30	20	40
Silty clay loam	0	40	25	35

Other Types

The texture is modified and the soil behaviour affected when a soil contains many stones or gravel, much organic matter or calcium carbonate (chalk or limestone). Stones and gravel can open up the soil. Thus an otherwise heavy soil can behave in the field as though lighter in texture if it is stony or gravelly. Much organic matter or calcium carbonate or both give the soil a smooth silky feel, as if it were silty but usually both types can be identified by their colour.

Gravelly soils are generally unsuitable for fruit as they have inherent defects: the lighter types are often too open in structure and are particularly subject to drought. Gravel pans or consolidated layers of gravel are also common on gravelly soils. These pans interfere with root development and can give rise to very bad drainage conditions. The only gravel soils which should be considered for fruit are those where there is no evidence of consolidation pans or bad drainage and where the texture of the soil between the gravel is at least as heavy as a loam.

Soil Structure and its Stability

Structure and its stability are important because of their influence on the cultural operations that can be undertaken on the site. Special precautions may be necessary to avoid damage to roots on soils of unstable structure, for example, silty loams and sandy clay loams. Grassing down may be necessary at a very early stage and the movement of machinery over the soil may have to be restricted as far as possible, and to take place when conditions are most suitable. This can be a serious handicap in orchard management. Sites should be rejected where structure is poor or planting should be delayed until conditions have been improved.

Good soil structure and stability of soils under adverse conditions are very important considerations in fruit production, though often neglected. It is not sufficient to start with the orchard in good condition; a watch must be kept on the effect of cultural treatments and operations on soil conditions. Deterioration in a good soil structure can very rapidly lead to extremely poor soil conditions such as bad drainage and anaerobic conditions in consolidated layers in the soil. Once good structure is destroyed it may take much effort and expense to re-build.

Good structure is often associated with moderate to high levels of organic matter in the top soil and usually with good drainage throughout the profile. Where organic matter has been allowed to decline to a very low level the heavier clayey soils tend to become cloddy and difficult to work and the instability in predominantly silty soils is aggravated.

Soil structure assessments are best made by studying soil profiles exposed by digging small pits to 3–4 ft deep in a few areas of the site. Before planting fruit, the soil conditions should be improved where necessary. Soil pans should be broken up when these are within about 20 in. of the surface and can be reached by subsoiler, buster or mole plough. Unstable types of soils such as the fine and very fine sandy soils and silty soils are stabilized to some extent by increasing their content of organic matter.

Evidence of good organic matter conditions is found in the even distribution of well rotted brown humus throughout the surface soil and where there is no sudden change in soil colour but a gradual transition down the profile from dark brown to paler brown. Adverse conditions are often indicated by the accumulation of a layer of black organic matter in a shallow band on the surface. This can arise from prolonged wet conditions, or in excessively dry situations or from acute acidity. Accumulation of these layers of organic matter follows the lack of activity under adverse conditions of organisms which break down organic matter to humus.

It is difficult to raise organic matter levels rapidly, especially in open sandy soils. A drastic change in management is usually needed such as laying the land down to a long ley or grassing down an orchard. The addition of farmyard manure and straw and the use of mulches help to maintain organic matter levels, but unless applied at very high rates will not raise levels to any extent except after a long period of use.

Presence of Physical Obstructions

Consolidated layers of soil or soil pans can partially or completely stop the downward development of roots and can also prevent upward and downward movement of water. Drought effects are therefore caused due to lack of root development and inability of roots to obtain water which may be present underneath the pan. In addition, excess water following heavy rain may accumulate above the pan and cause temporary bad drainage conditions even in light sandy soils.

Soil Drainage

Most fruit crops require good drainage conditions since waterlogging results in the death of roots. It is generally unwise to choose inadequately drained land for fruit but where defects are only slight it is often possible to drain the land satisfactorily before planting. Adverse drainage conditions can result from high water tables, springs, and slowly permeable subsoils which may be a natural characteristic of the soil or due to damage caused by cultural operations. Soils subject to bad drainage include those which pack tightly such as the very fine sandy or silty types. Clayey soils, particularly eroded clays, may be badly drained but there are many clay soils which have good structure and have satisfactory drainage. In contrast, drainage is usually rapid in open, coarse sandy soils and crops on these soils can suffer from lack of moisture.

In the assessment of drainage conditions the main factors which have to be considered are the texture and structure of the soil, its depth and the colour of the soil throughout the profile. In diagnostic work, the examination of soil colours can provide valuable information on drainage conditions. The interpretation of soil colours requires experience and knowledge of the factors which affect colours, such as the influence of organic matter, wetness, lack of aeration, fluctuations in wet and dry conditions, acidity effects and the inherent colour of the parent material. Nevertheless soil colours provide to the experienced eye abundant information for assessing suitability of soils for different crops.

In general, uniformity of colour without abrupt or marked changes down the profile indicates a well drained soil. A dark brown surface soil which gradually changes to warm brown colour extending two or three feet indicates a soil suitable for extensive rooting. Grey, greyish-blue, rusts and mottled grey-rusts in the profile give indications of adverse drainage conditions. Grey and greyish-blue colours indicate impeded drainage but mixtures of greys and browns and grey with rust markings indicate fluctuations of poor drainage and aerated conditions. Past bad drainage is often indicated by rusts, small rusty or black concretions, and by rusty streaks where no grey occurs. If grey colours are also present then drainage is still faulty.

Experience in the interpretation of soil colours is necessary as there are some conditions which can mislead the unwary. Thus grey colours do not occur in the Old Red Sandstone and Keuper Marl soils to indicate bad drainage: the main change in colour is from a dark to a pale red. Another example is the effect of ground water and spring water moving through soils. These can remove much iron, which is the source of the diagnostic colours in soils. The soil may have a bleached appearance which remains after the moving water has disappeared but this does not necessarily indicate poor drainage. If the soil is then subject to bad drainage no greys or blue colours appear and there are thus no colour indications of adverse drainage.

Black types of organic matter often indicate wetness or past wetness, and in sandy soils present or past acidity. Extremely dry conditions such as occur on a shallow chalk downland soil can lead to the accumulation of a surface layer of black organic matter. Bleached sand is usually an indication that the soil has been subject to extreme acidity at some stage in its history.

There is some variation in the tolerance of the various classes of fruit to different drainage conditions as shown in Table 4.

TABLE 4

Tolerance of Fruit to Various Soil Drainage Conditions

Type of Fruit	Drainage of the Soil
Sweet Cherries, Raspberries .	Require well drained soils. Will not tolerate inadequate drainage.
Dessert Apples, Gooseberries .	Will tolerate slight drainage impedance below about 18 in., except the gooseberry variety Leveller which requires a well drained soil.
Culinary Apples, Plums, Pears Red currants, Blackberries, Loganberries	More tolerant than above two classes. Will tolerate drainage impedance below about 18 in.
Black currants . . .	Can withstand imperfect drainage when well supplied with nutrients.
Strawberries . . .	If the top 12 in. are well drained then they can tolerate some bad drainage underneath.

The pH and Lime Status of the Soil

The pH of the soil is a measure of its acidity or alkalinity. A pH of 7·0 is neutral; soils with a pH above this are alkaline and below this value they are acid. The pH of soils in this country ranges from about 4·5 (extremely acid) to 8·3. Soils of moderate to slight acidity, pH 5·5–6·5, are most suitable for the majority of fruit crops. Excessively acid conditions, pH 5·0 and below, result in root damage due to toxicity of manganese and aluminium which are excessively available at low pH values (see page 29). Acid soils should be limed to a pH of 6·0 before planting except in the case of blueberries, which require a pH of 4·5–5·5. Black currants are more sensitive to acidity than most fruit crops and a pH of 6·5 is advisable. A pH of 6·5 is also desirable for new plantings of dessert apples in soil strips kept weed free by herbicides. The soil in these strips soon becomes acid and this can lead to manganese toxicity. On calcareous soils on the other hand manganese and iron deficiencies can occur separately or together. The fruit crops most susceptible to iron deficiency are pears, raspberries, and strawberries. Apples are somewhat less susceptible.

The Nutrient Status

As deficiencies of nutrients can be corrected by suitable applications of manures, fertilizers and foliar sprays, the nutrient levels in the soil, both in the surface and at lower depths, need not be regarded as limiting the value of the land for fruit.

Freedom from Injurious Materials

Crops can be damaged very seriously by an excess of some elements. The most serious toxicities are often those where the excess of an element or elements has been added inadvertently. Materials contaminated by industrial wastes, such as sewage sludge from an industrial area, can sometimes add sufficient zinc, copper, nickel and other elements to damage crops and eventually kill them. These types of toxicity are extremely serious as the excesses persist for very many years and so *materials which are contaminated with metals should never be used.*

Soil Requirements of Different Fruits

Strawberries

A minimum depth of soil of 15 in. is desirable, but slightly less depth of well drained soil is tolerable if rooting is possible in the underlying soil. The most suitable textures are loams and fine or very fine sandy loams. Where drainage is good main crops are sometimes grown satisfactorily on clay loams.

Iron deficiency is common in strawberries grown on calcareous soils.

Raspberries

Heavy soils in the clay class are generally unsuitable for raspberries which prefer light to medium textured soils. A minimum depth of soil of 18 in. is necessary and this should be well drained. Raspberries are unsuccessful in poor drainage conditions which give rise to die-back of shoots and death of canes.

The raspberry varieties Norfolk Giant and Malling Promise are very sensitive to any adverse soil conditions, particularly wetness, but Malling Jewel will often succeed where there are slight defects in soil drainage. There is as yet insufficient experience with the newer varieties to assess their reaction to soil conditions.

Raspberries are very susceptible to iron and manganese deficiencies when grown on calcareous soils.

Blackberries and Loganberries

These crops can be grown successfully on a wide range of soil textures but in general the lighter and medium textured soils with good water holding properties are preferable to clay loams. Wetness of clay soils in winter can cause the canes to die out but they are less sensitive than raspberries to defects in soil drainage.

Blackberries are fairly susceptible to manganese and iron deficiencies.

Black Currants

These are grown successfully on a wide range of soil textures. As the best production occurs on well-grown bushes, a moderately deep soil of 18–21 in. with good water holding capacity is desirable. Black currants are more tolerant than raspberries and red currants of bad soil drainage.

Black currants do not tolerate acidity and prefer soils with a pH in the range of 6·0–6·5. The variety Wellington XXX is less sensitive than most other varieties to soil conditions. Baldwin in contrast requires good growing conditions otherwise the bush remains small. Black currants are less susceptible than most fruit crops to manganese and iron deficiencies.

Red Currants

Red currants can be grown on a wide range of soil textures from sandy loams to sandy clay loams provided drainage of the soil is good. In general, the lighter textures are preferable provided they are not too susceptible to drought. Red are similar to black currants in their requirement for moderately deep soils which are slightly acid; they are less tolerant of adverse soil drainage and are slightly more susceptible than black currants to iron deficiency.

Gooseberries

Gooseberries require good soil conditions. At least 18 in. of well-drained loam is desirable. Sandy loams subject to drought and heavy shallow soils are unsuitable. The variety Leveller is more sensitive than other varieties to adverse soil conditions and does badly on soils with impeded drainage.

Gooseberries are very resistant to both manganese and iron deficiencies when grown on calcareous soils.

Blueberries

Blueberries will not tolerate alkaline conditions and succeed only on strongly acid soils. They can withstand slightly impeded drainage but will not grow satisfactorily in soils susceptible to drought in summer.

Apples

Dessert apples are grown successfully on a wide range of soils from sandy loams to clay loams.

The rootstocks used for apples show some differences in their reaction to particular soil conditions. For example, trees on the dwarfing rootstock M.9 have shown a wide degree of tolerance of soil conditions. Trees on this stock have succeeded on shallow inadequately drained clay soils. On the other hand trees on MM.104 have proved especially sensitive to impeded drainage and this stock should not be used under such conditions. Trees on MM.111 have shown considerable resistance to drought and are better in this respect than trees on M.2, a stock of similar vigour. MM.111 may, therefore, be a good choice for light sandy soils of poor moisture holding capacity. On such soils also, trees on the vigorous stocks such as M.25 may prove useful in maintaining adequate growth whereas on more fertile soils they may produce too large a tree.

Differences between the behaviour of two comparable stocks are often related to soil conditions. For example, trees on MM.106 grown on a good loam have been about equal in size to those on M.7, whereas on a light sand the trees on MM.106 have been much smaller.

In general, shallow light soils subject to extreme dryness and heavy soils with bad drainage should not be planted with dessert apples. The ideal soil is a well-drained loam. A minimum depth of soil of approximately 24 in. is necessary, particularly on light soils, but if some rooting is possible in the underlying material 18 in. of a well-drained soil can be suitable.

Culinary apples are less sensitive to inadequate drainage than dessert apples and succeed on a wider range of soil textures.

Apples grown on calcareous soils are susceptible to both manganese and iron deficiencies.

Pears

Pears are grown successfully on a fairly wide range of soil textures from sandy loams to clay loams but do poorly on dry light types. Compared with apples they are less tolerant of drought and rather more tolerant of impeded drainage. A depth of about 24 in. of soil is desirable but slightly less may be suitable provided some rooting is possible in the underlying material and that this is also a source of moisture.

Pears on quince rootstocks are very susceptible to iron deficiency.

Plums

Plums are successful on the heavier soils of the loam to clay loam textures. Branches of trees are brittle and break easily where moisture supply is inadequate as, for example, on coarse sandy loams. Plums are more tolerant than dessert apples of adverse drainage but will not tolerate waterlogging. They are similar to pears in requiring a moderately deep soil. The minimum depth is approximately 24 in. depending on the underlying material.

Of the varieties, Early Rivers is one of the more tolerant of adverse drainage. Czar and Monarch are better than Victoria on dry soils.

Plums are fairly susceptible to iron deficiency.

Cherries

Cherries are grown successfully on a wide range of soil textures from sandy loams to clay loams. They require very deep soils and do not tolerate poor drainage conditions. The deep sandy loams to loams are often more suitable than the heavier soils. Cherries require a depth of soil of at least 30–36 in. with a favourable water supply during the summer months. Of the sweet cherries, the variety Early Rivers tolerates slight defects in soil conditions better than other varieties, such as Bigarraeu Napoleon which does poorly if there are any defects in depth of soil and its drainage conditions. Sour cherries tolerate less depth of soil than sweet varieties.

Cherries are not as susceptible as pears and plums to iron deficiency, but a deficiency of iron is not uncommon when cherries are grown on calcareous soils.

Vines

Vines will grow well on soils varying widely in depth, texture and drainage. They are less susceptible to drought or adverse drainage conditions than most fruit crops but for optimum yield, although not necessarily optimum quality of wine, a deep, well-drained and moisture retentive soil is required. Since considerations of site and aspect are so important the choice of soil cannot be given first priority. Shallow or coarse sandy soils are acceptable provided roots can penetrate to three feet or more and naturally poorly drained soils are adequate provided artificial drainage is very intensive.

SOME OF THE MAIN FRUIT AREAS AND THEIR SOILS

The Vale of Evesham

The Evesham Series

This is the most widespread and important soil series of the area and is a clay loam over clay. The soils are formed mainly from the Lower Lias Clay which is highly calcareous. They are moderately well drained in the surface layers but drainage is slow. Structural conditions are often good in the Evesham clay loam which cracks and fractures under the influence of wetting and drying, and of frost. Excellent tilths can be obtained on these soils. Orchards, mainly of apples, are widely distributed on these soils which provide conditions for vigorous growth of less exacting varieties of culinary apples. Plum and pears are grown successfully and the quality of plums is excellent. Most soft fruits, except raspberries, grow satisfactorily. The soils however, are not well enough drained for sweet cherries.

Soils of the Avon Terraces

These soils are generally very sandy. They are often drought susceptible and inherently deficient in potash and magnesium.

Pershore Series

Soils of the Pershore series are stony loamy sands or fine sandy loams over sand. They overlie Lias clay at depth and the drainage is impeded and wet spots occur in winter. When tile drained the fine sandy loams are satisfactory fruit soils but the sands are too drought prone.

Wick Series

This soil is a deep well drained sandy loam with some pebbles. Although not excessively drought susceptible, irrigation is desirable for fruit.

Newport Series

Soils of this extensive series consist of well drained loamy sand or sandy loam over sand often with pebbles. They are too drought susceptible to be satisfactory for fruit unless irrigated.

Dunnington Heath Series

This series is similar to the Pershore series usually consisting of loamy sand over sand with impeded drainage since they overlie Keuper Marl at depth. When tile drained some of these soils are suitable for fruit but the lightest types are very drought prone so that irrigation is needed.

Herefordshire Fruit Soils

Bromyard Series

This widespread soil consists of a reddish-brown silt loam overlying marl. Most fruit crops grow successfully on the deep, well-weathered soils but poor growth and crop failures may occur on the shallow, eroded and poorly weathered soils. Small wet areas of soil are common where bands of marl or unweathered material near the surface cause impervious layers holding up water movement especially on slopes. These soils are susceptible to erosion on the steeper slopes. Although generally well drained the drainage is slow and careful management is necessary.

Eardiston Series

Soils of the Eardiston series are deep reddish-brown fine sandy loams and are very good fruit soils. Like the Bromyard series they occur on valley slopes and can be prone to erosion.

Ross Series

This soil is similar to the Eardiston series but the sand is considerably more coarse and it is a coarse sandy loam. When deep over sandstone it is a good fruit soil but when shallow it suffers from drought and trees may be stunted in patches. The Ross soils are less well supplied with potash and magnesium than the Bromyard and Eardiston soils.

The Lower Greensand of Kent

The soils on the Lower Greensand of Kent vary in character depending upon the particular division from which they have been derived. In general they are suitable for growing fruit trees but in some areas extremes of texture with their related moisture characteristics, for example, coarse texture and low available water capacity; fine texture and poor drainage and aeration, and excessive shallowness regardless of texture render the soils unsuitable for fruit.

The greatest area of suitable soils is to be found in the Hythe Beds division and much of the dessert varieties of top and soft fruit lies on this formation.

These soils are mainly deep, fine sandy and silty loams but slightly heavier soils may also be found to be satisfactory if they are well drained. Shallow soils overlying Ragstone occur in places and these are not suitable for fruit.

The soils derived from the Sandgate Beds can be satisfactory for fruit but often in this formation seams of clay and sandy clay can impair the drainage and in general these soils are less suitable than those derived from the Hythe Beds.

The Folkestone Sand has produced rather coarsely textured soils having low available water capacities which are unsuitable for fruit but in places where there is an admixture of windblown drift such as brickearth the water retentive properties of the soil have been improved and such soils are more suitable than the majority on this formation.

Outcrops of Atherfield Clay are small in area and are generally avoided when fruit is planted.

The following soil series are of value for fruit:

Malling Series

These soils are derived from Hythe Beds drift. They are deep well drained loams to clay loams with moderate available water capacity overlying Ragstone interbedded with loamy Hassock. They are suitable for all classes of fruit.

Barming Series

These soils have been formed on loamy drift from Brickearth and Folkestone Beds over sandy Folkestone Beds. The drainage is good and the available water capacity is moderate to large. Suitable for all classes of fruit.

Ditton Series

These soils are similar to the Barming Series but they are slightly less well drained below about 15 inches. The better drained phases of this series are suitable for all types of fruit, but those more poorly drained will require artificial draining.

Hamble Series

These soils are deep well drained silty loams and have large available water capacities. They are derived from Brickearths and are of widespread occurrence, overlying many of the geological formations in the south-east. They are suitable for all classes of fruit.

The Hastings Beds of Kent and Sussex

One of the features of the area is the frequent variation in soil type, even within an orchard. Very careful soil examinations are therefore required before planting fruit.

Pembury Series

These soils are well-drained fine sandy loams to loams in texture, from 16–24 inches in depth. There are some shallower than this and these are unsuitable for fruit. The moderately deep to deep Pembury soils are excellent for all kinds of fruit but especially for dessert apples.

Curtisden Series

These soils are somewhat similar in texture and depth to the Pembury Series but drainage is impeded although waterlogging does not occur. Soft fruits of the stronger growing varieties are successful but dessert gooseberries are not satisfactory. Vigorous varieties of culinary apples and plums are successful though growth is not as vigorous as on the Pembury Series.

Ladham Series

Soils in this series are developed on Valley Brickearth and range from very fine sandy loams to loams in texture. They are deep and well drained and are excellent for fruit, especially for cherries, pears and soft fruits.

Teise Series

The soils in this series are derived from drift and colluvial material in the High Weald and show evidence of impeded drainage. Springs are the main source of drainage problems. The soils are mainly very fine sandy loams to loams. Where there is 18 in. or more of depth and spring water seeps below, there is strong growth of vigorous varieties of most tree fruits, and of soft fruits. Dessert apples and especially cherries do badly in these conditions.

The North Kent Fruit Belt

This is a strip of land on the lower dip slope of the North Down between Canterbury and Sittingbourne. The soils are mainly deep, well drained silty or fine sandy loams derived from the Brickearth and Head deposits and the Tertiary deposits, Thanet Sand and Woolwich Beds.

Three soil series developed on the Brickearth are suitable for fruit:

Hamble Series. This Brickearth soil is described above.

Hook Series. These soils are similar to the Hamble Series except that signs of impeded drainage may occur below about 15 inches. They are suitable for most types of fruit but, unless artificially drained, only those better drained should be used for growing dessert apples and cherries.

Parkgate Series. This is a series developed from Brickearth in which the drainage is impeded as a result of underlying impervious formations such as London Clay. The soils require intensive drainage to make them suitable for fruit.

Coombe Series

These are soils of variable depths derived from Coombe deposits and usually contain a good deal of silty drift such as Brickearth. Their drainage is good and the deeper phases in which it is possible for the bulk of roots to be free from the influence of the underlying chalk are suitable for most types of fruit. Their available water capacity is high. There is, however, a risk of trees being affected by iron deficiency though symptoms are not usually severe except in shallow areas.

Bursledon Series

Soils of this series are formed in loamy deposits over and partly from Thanet and Woolwich Beds. They are sandy loams at the surface becoming heavier with increasing depth and though like the Hook series, which they resemble, the drainage is imperfect at depth, they can be made suitable for fruit by tile drainage.

Woodnesborough Series

This series is developed on and mainly from the Woolwich Beds but with some admixture of Brickearth. The soils are similar to the Barming series. They are sandy loams and tend to be a little drought susceptible except where the sand is fine. With irrigation most of the soils are very satisfactory for fruit.

Berkshire Fruit Soils

Harwell Series

This soil occurs in a narrow band of upper Greensand on the lower slopes of the Berkshire Downs. It is usually a very fine sandy loam over silt loam overlying weathered malmstone and is excellent for all fruit crops. The malmstone is occasionally calcareous. As well as being usually well drained it has a high available water capacity. The potassium status is naturally very high. Apart from needing drainage in some areas the only problem is iron deficiency in pears in alkaline areas.

Kingston and Fyfield Series

These soils occur on the Corallian Sands. They are mainly deep sandy loams but the texture includes areas of loamy sand and sandy clay loam. Growth is not vigorous due to limited available water capacity. The pillar system has been used successfully for growing dessert apples. Zinc deficiency has been diagnosed in one orchard.

Hampshire Fruit Soils

Hamble Series

This Brickearth soil is described in the section on Kent soils.

Icknield and Andover Series

Although shallow and calcareous these light chalky soils are cropped with dessert apples, strawberries and vines. Iron and manganese deficiencies can be problems.

West Cambridgeshire Fruit Growing Area

There are two main groups of soils in this area: a larger area to the north of Cambridge where a range of soils is derived from peat, heavy alluvium, gravels, Greensand and clays, and a smaller group to the south of Cambridge in the district of Melbourn where the soils are derived mainly from the chalk and chalk marls. The main fruit crops are plums and apples, grown mainly on loamy soils derived from the local loams and marls. They are occasionally grown on light gravelly, sandy and chalky soils.

Strawberries are the main soft fruit crop and are found, as in the Wisbech area, mainly on the heavier soils but are seldom grown on soils derived from Chalk Marl and the Chalk. The soils derived from river terrace gravels and loams are divided into three main soil series: Landbeach, Milton and St Lawrence.

Landbeach Series

The soils are mainly sandy clay loams with some stones. Drainage is usually good but the water table is high in some areas necessitating tile drainage. Trees grow satisfactorily in the young stage but quickly deteriorate and lack vigour. Apple and plum trees remain small and leaf 'scorch' and die-back can occur on the drier types due to drought.

St Lawrence Series

These soils are mainly clay loams or sandy clay loams with some stones, over grey clay. Drainage is often slightly impeded and can be bad locally. Good crops of Victoria and Czar plums are produced where drainage conditions are reasonable.

The highly calcareous fruit area of Melbourn is mainly situated on two soil series: Wantage and Burwell. Although far from ideal fruit soils due to their chalky nature, plums and dessert apples have long been grown successfully in spite of problems with iron deficiency.

Wantage Series

These well drained soils occur on the lower slopes over chalk or chalk marl. They consist of about 18 in. of silty loam to silty clay loam over the chalk or marl.

Burwell Series

The Burwell series is similar to the Wantage series except that it is lower lying and less well drained. Tile drainage is needed in some areas.

The Wisbech Fruit Area

The soils of the area are derived mainly from silt alluvium but peat alluvium also occurs in areas near Wisbech. The main soil types on which fruit is found are the fine and very fine sandy loams, silty and silt loams, also silty clay loams and some organic types where the soils are derived from a mixture of peat and mineral alluvium. The soils are usually deep and have a ready supply of water at depth. The main tree fruits grown are culinary apples, but there is a considerable acreage of dessert apples also and of plums. Of the bush fruit crops, gooseberries are the most commonly planted. Strawberries are a very important crop in the area. In contrast to the growing of strawberries on light land in south-east England the largest strawberry areas are found on the heaviest soils of the Wisbech district (silty loams to silty clay loams) and the fruit is of the best quality on these heavy types. Strawberries are also grown and do well on the lighter silt soils which are equally moisture retentive although better drained. The main difficulty encountered with strawberries on heavy silts is occasional damage by wetness in winter. Runners sometimes find difficulty in rooting into surface soil which has crusted in a dry period after a previous heavy rain.

On the lighter silty soils, apple trees become large and strong growing and usually make stout, long, annual growth. The fruit is also largest on light silts but colour of fruit is less pronounced. The heavier silty soils tend to produce smaller trees. The soil series have not so far been identified in this area.

Norfolk Fruit Soils

Except for that part of the Wisbech area which lies in Norfolk most of the fruit grown in Norfolk consists of black currants, gooseberries and raspberries. Many of the Norfolk soils are suitable but some of the lighter ones are drought susceptible.

Sheringham Series

These soils are deep, well drained, very fine sandy loams and are excellent for all fruit crops although they occur rather far north for tree fruit.

Hall Series

Soils of this series are well drained fine sandy loams over sand. They are suitable for soft fruits although a little susceptible to drought.

Aldeby Series

This imperfectly drained soil consists of about two feet of sandy loam overlying sand overlying clay. When tile drained it is a good fruit soil and several successful dessert apple orchards are on this series.

Suffolk Fruit Soils

Suffolk is noted for top fruit particularly dessert apples which have a good reputation for quality. However, many of the orchards are situated on difficult heavy soils which require intensive drainage.

Ashley Series

This is one of the best Suffolk fruit soils and consists of sandy loam or sandy clay loam overlying chalky boulder clay. It is moderately well drained but tile drainage is desirable to avoid root loss during exceptionally wet seasons.

Beccles Series

Most orchards are situated on this series which is imperfectly drained and often not very suitable for fruit. At best the soils are similar in texture to the Ashley series but the drainage may be slightly inferior. When intensively tile drained at spacings of about half a chain they can be satisfactory. However, many of the Beccles soils are heavy sandy clay loams and difficult to drain effectively because mole drains are unstable. Trees tend to suffer in wet seasons even with intensive tile drainage and long term average yields are not very good.

Ragdale Series

These soils consist of about 12 inches of heavy sandy clay loam overlying chalky boulder clay and are often calcareous in the topsoil. Although heavy and requiring careful management they are more suitable for dessert apples than many of the Beccles series. The underlying chalky boulder clay builds up a good stable structure when intensively drained, and mole drains are effective.

Essex Fruit Soils

A wide range of soils of varying suitability are used for fruit in Essex ranging from sands and gravels over London Clay to deep brickearths and chalky boulder clays. The Ashley series occurs in parts of Essex as well as Suffolk.

Hanslope Series

This chalky boulder clay soil consists of calcareous loam or clay loam over calcareous clay with many chalk stones. It is naturally moderately or imperfectly drained but when tile and mole drained it becomes virtually well drained. It is a good soil for dessert apples, black currants and gooseberries but iron deficiency can be a problem in pears.

Bengeo Series

This rather variable soil consists of flinty loam or sandy clay loam overlying flinty clay loam and gravel at depth. It is well or moderately well drained and suitable for most fruit crops. Tile drainage is necessary in some areas.

Hamble Series

This Brickearth soil is described in the section on Kent soils. In Essex the texture is usually very fine sandy loam.

Hook Series

Similar to the Hamble series but imperfectly drained. With tile drains at one chain spacing it is as satisfactory as the Hamble.

Wickham Series

This series consists of imperfectly drained sandy loam, loam or clay loam over London Clay. The lightest and deepest types are moderately suitable for fruit but difficult to drain effectively and subject to wet spots.

Somerset and Gloucestershire Fruit Soils

Atrim Series

The Atrim series on the slopes of the upper and middle Lias siltstones is a well drained fine sandy loam over micaceous silt loam. It is usually deep and quite satisfactory for fruit.

Martock Series

The Martock series is similar to the Atrim series but is affected by springs and needs tile drainage in wet spots.

Worcester Series

These reddish-brown soils are silt loams over Keuper Marl. They are usually deep and moisture retentive but are slow draining and liable to soil compaction. Occasional subsoiling is practised in many orchards and tile drainage is necessary at some sites. In some areas the soil is shallow over the marl so that in consequence trees suffer from poor root development and drought.

Mineral Requirements of Fruit

Fruit plants are perennials of varying growth habits and include trees, bushes, canes and the herbaceous strawberry. Fruits are grown over a wide area and so are subject to a range of climatic conditions, particularly the amount of rainfall, summer temperatures and sunshine, each of which may modify the vegetative growth and fruiting characteristics. Perennial plants live for a number of years and for fruit trees this may be 20 to 50 years or longer. The nutritional requirements will vary over the period. Growth of both root and shoot systems will have an effect on the nutrient requirements. Unbalanced manuring over a long span may result in the accumulation of some nutrients to undesirable levels. For example, the effect of excess potassium in reducing the uptake of magnesium is well known. Moreover, if arable cultivation is continued for a long time without regard to organic matter status, soil conditions may become unfavourable for good root growth and nutrient absorption.

The quality of the crop and in particular its appearance and its storage properties as well as yield must be carefully considered when assessing the effect of manures. These commercial considerations will become even more important with the increasing emphasis now placed on fruit quality. Furthermore, manurial practice will be influenced by cultural systems and by pruning methods.

Mineral Nutrition

It has been established by research that in addition to carbon, hydrogen and oxygen, higher plants require for healthy growth through a complete life cycle at least 12 elements, generally termed the essential mineral nutrients. All the 12 elements are essential for full growth but for convenience they are divided into two groups based on the amounts required.

Major or macro nutrients	Minor, micro or trace nutrients
Nitrogen	Iron
Phosphorus	Manganese
Potassium	Boron
Calcium	Copper
Magnesium	Zinc
Sulphur	Molybdenum

Unless these essential elements are available in the soil or supplied artificially in adequate quantities at a stage when the crop requires them, satisfactory yields will not be obtained. Other elements which are present in soil may be absorbed by roots and pass into various organs of plants. The uptake of excessive amounts of some elements produces no obvious effects, but others can be injurious and produce unhealthy growth. For example excess zinc will upset growth, and similarly excessive amounts of manganese can be taken up from acid soils.

The effects of deficiencies or excesses of nutrients are shown in various ways by plants and especially by the leaves. Deficiencies of certain elements or sometimes the effects of pests and diseases cause leaves to turn yellow and in extreme cases, almost white. This change in the natural colour of leaves is termed chlorosis. Often fruit trees growing in soils of a high lime status look yellow and these are spoken of as suffering from lime-induced chlorosis. Sometimes part of the leaf tissue dies resulting in brown areas in the leaf. This symptom is called necrosis.

In the field, deficiency symptoms are often confused and sometimes obscured by the effects of weather conditions, spray damage, insect attack or virus infection but in general nutrient disorders produce a *regular* pattern of symptoms. A system of visual diagnosis of deficiency symptoms has been developed which is very useful to experienced observers, but often field diagnoses need to be supported by laboratory examinations.

Factors Influencing Nutrition

Growth of Roots

Mineral nutrients are taken up by roots mainly from the soil solution. This absorption process can only be carried on by active living roots. The uptake of plant foods depends on many soil factors. The soil must contain sufficient moisture for growth but yet must be well enough aerated for the roots to take in oxygen for respiration; this means that good fruit soils must possess a good structure and the husbandry system must be one that will maintain a good soil structure.

Some cultural practices can reduce root growth by damaging soil structure. The passage of tractors and implements over the soil when it is too wet can result in the formation of pans and compacted layers near the surface. The depth of soil is important and for each type of fruit a minimum depth of well-drained rooting medium is needed for full cropping performance. This depth will vary somewhat with the climate and location.

Weather Factors

Weather conditions also have an effect. For example, in dry seasons the effects of acidity are more apparent, whilst wet conditions often make magnesium deficiency worse. Mineral deficiencies sometimes occur in very wet seasons because nutrients can be leached from the leaves producing a paleness around the margins or occasionally necrotic spotting.

Nutrient Interactions

Some nutrients interact with one another; excessive amounts of one nutrient may reduce the intake of another and *vice versa*. Perhaps the best known example of this is the effect of excess potassium in aggravating magnesium deficiency. Calcium is also important because of its effect along with other so-called bases, potassium and magnesium, on the acidity (pH) of the soil. In strongly acid soils, that is, soils with a pH value below 5·5, aluminium and manganese may be present in toxic quantities. On the other hand, excess calcium can make soils alkaline (pH value above 7·0) and this can cause trouble by reducing the intake of iron and manganese. The overall effect is complex and whilst soils should not be too acid, over-liming of fruit soils must be avoided.

Fertilizer Effects

The type of fertilizer used produces effects in addition to those of nutrition. The continuous use of sulphate of ammonia will make a soil more acid and where an orchard or plantation is situated on a soil which is naturally acid it is advisable to use a fertilizer of the ammonium nitrate/lime type as a source of nitrogen. When nitrogen is supplied in the form of nitrate, magnesium deficiency tends to be alleviated but nitrogen in the ammoniacal form makes the deficiency more acute. If muriate of potash is used to supply potassium the chloride may cause damage in certain circumstances, especially to soft and cane fruits.

Soil Factors

Soils can be classified mechanically into fractions of various sizes. In a conventional mechanical analysis the amounts in four size groups are usually measured, these are: coarse sand, fine sand, silt and clay and the proportion of these in any one soil determines the texture. Particles in the first three size ranges are essentially unaltered minerals that have been ground down by various weathering processes but much of the clay sized particles have been produced by chemical changes during weathering. They are very small and so the surface area for a given weight is very large. Some of the soil organic matter has similar surface properties and the intimate mixture of these two which exists in most soils is known as the clay-humus complex.

The chemical substances known technically as bases or cations, are held on the surface of the clay-humus complex with varying degrees of firmness. In the soils of this country the main bases are calcium, potassium and magnesium. If there is a large amount of any one base present in solution it will displace other bases from the surface of the complex and these appear in solution. In this way the composition of a solution can be changed by pouring it through a sample of soil held on a filter. This process is known as base exchange and the base exchange capacity is an important property of a soil. In general the heavier the soil texture and the greater the amount of organic matter, the higher the base exchange capacity. In the field there is constant exchange between bases in the soil solution and those on the surface of the soil complex which acts as a reservoir. When certain fertilizers are applied they dissolve first in the soil solution and then are adsorbed by the complex. There the firmness of binding depends on the mineral make-up of the clay and on the type of organic matter. The roots absorb nutrients from the soil solution and in some soils, at a time when growth is very active, the rate of release from the complex may not keep pace with absorption by the vigorous roots and a temporary deficiency may arise.

Potassium is usually held fairly firmly by the exchange complex and is not easily lost by drainage except in very light textured soils. Magnesium is more easily leached and sometimes excessive amounts of potassium (or calcium) will displace magnesium from the complex and increase the rate of loss. On the other hand phosphorus moves only slowly in the soil and thus little if any is leached out of the soil.

Nitrogen differs from other nutrients because it is produced in the soil mainly from organic matter, whereas all the other nutrients can be produced by the weathering of the inorganic soil minerals. Of course, organic matter in the soil is formed from dead organisms and plants which contained a range of nutrients and these are released when organic matter breaks down. Various micro-organisms in the soil take the nitrogen in the organic matter through a chain of processes and it is eventually released as nitrate. These micro-organisms living in the soil need nitrogen for their own life and growth and sometimes they remove nitrogen from the soil, to break down organic matter, and thus starve the crop. Similarly, one crop may use nitrogen and cause another crop to be short as with the well known effect of nitrogen shortage in trees when an orchard is grassed down. Nitrate nitrogen is not held firmly in the soil and is therefore liable to loss by leaching, particularly in light textured soils. Where the texture is heavier leaching is less and it is only when rainfall is heavy and prolonged that nitrate is removed. Often nitrate, although washed down from the surface soil, will still be present lower down in the root zone.

The whole environment of the actively growing plant root, solid soil particles, soil moisture and soil air must be regarded as a constantly changing system. The roots take in oxygen and release carbon dioxide. This alters the acidity of the soil solution which will in turn affect the amount of nutrients coming into solution from the solid particles.

In dry conditions the soil solution will be more concentrated whilst when the soil is waterlogged oxygen will be in short supply. Roots absorb nutrients mainly from the soil solution and the roots of some plants are more effective in doing this than others. This removal causes a change in the balance in the solution and more nutrients will come into solution. In some soils this latter process may be rather slow and so insufficient nutrients will be present in solution and perhaps growth may be retarded. When fertilizers are applied some of the nutrient may go on to the solid particles and become fixed and some may be lost by leaching. A proportion of the nutrient added as fertilizer will be held quite loosely on the soil surface and so will be easily available to replenish that in solution.

Effects of Deficiencies and Excesses

NITROGEN

Deficiency symptoms. Shoot growth is restricted and shoots are thin; the bark of trees may be reddish or yellowish-brown in colour; leaves are small, tend to be sparse and are pale green in the early stages, later developing tints of yellow, red or purple; the foliage looks thin and there are often many relatively bare shoots, due to the death of lateral buds or their failure to develop (in strawberries this shows as a failure to develop new crowns); leaf fall is early; fruit bud formation is reduced and blossoms are weak; fruit yields are reduced, due to decreases in number and size of fruit which may be highly coloured, with pale ground colours in tree fruits, such as the apple or pear; the flesh of the fruit is generally hard and may be sweet, but in instances of acute deficiency may be woody and lack flavour; with moderate deficiency in apples and pears, dessert quality may be high, storage quality is generally good. Severe symptoms are often seen in orchards just after grassing down. In neglected orchards and in those where soil conditions are poor the main symptom is often that of nitrogen deficiency.

Excess symptoms. Where other elements are present in adequate amounts, trees and bush fruits are liable to be over-vegetative, foliage is luxuriant and the leaves large and dark green in colour; fruit yields may be poor and the individual fruits large and poorly coloured; the fruits are generally poor in eating quality and in storage properties, being highly susceptible to physiological breakdown and to fungal rots. In soils low in potassium, excess nitrogen may accentuate the potassium deficiency sometimes with severe effect on the trees. On the other hand high nitrogen often reduces the effect of magnesium deficiency.

PHOSPHORUS

Deficiency symptoms. Effects of phosphorus deficiency on growth have rarely been seen in the field in this country but the general effects, as produced in sand culture experiments with a range of fruit crops, resemble to a considerable extent those produced by nitrogen deficiency. Thus shoot growth is much restricted and shoots are thin; foliage is sparse and leaves small; leaf and blossom buds are few and there is much bare wood; the time of opening of the blossoms is delayed; defoliation is early. It is known that phosphorus is important for black currants and adequate amounts are necessary for good clover growth of a sward after grassing down.

In recent years, a low level of phosphorus in apples has been associated with poor storage qualities, in particular, with some forms of senescent breakdown in Bramley's Seedling apples.

Excess symptoms. So far, effects have not been observed in commercial fruit crops in this country but excess would make zinc deficiency more serious.

POTASSIUM

Deficiency symptoms. Leaves are small, often bluish-green in colour and they develop a marginal scorch which is a very characteristic symptom. Shoots are thin and growth is poor. These symptoms increase as the deficiency becomes worse and shoots or even the whole branch may die back. Under conditions of slight deficiency a great deal of blossom is produced but severe deficiency reduces flowering. Fruits are often small and rather flavourless. They ripen unevenly and yields are much reduced. In apples, many fruits are liable to fall from the period of 'June drop' onwards, and often few fruits remain on the trees at harvest time even when the original set was fairly heavy. Apples are most subject to low temperature breakdown in refrigerated storage. The amount of potassium in a soil depends on soil type and on past cropping and manuring. Soils derived from some geological formations, for example Upper Greensand and Chalky Boulder Clay, are naturally rich in potassium. Potassium deficiency is rarely found in these soils but it is common on light textured and shallow soils.

Excess symptoms. In the early years of a fruit plantation it is often important to manure generously with potassium to build up supplies in the soil and ensure good growth. Later on dressings may be reduced to avoid the accumulation of an excess which could reduce the intake of magnesium thus causing severe effects particularly in apples. High potassium may also increase the incidence of bitter pit and other storage disorders in apples.

CALCIUM

Deficiency symptoms. An absolute deficiency of calcium is unlikely to occur in the field except possibly on strongly acid, light sand soils, but symptoms of die-back and marginal leaf scorch have been described from sand culture experiments.

Recent work on apple storage has emphasized the need for adequate levels of calcium in the fruit; low levels of calcium being associated with a high incidence of bitter pit which shows as brown corky areas within the cortex, sometimes accompanied by small depressions visible on the surface of the apples.

Soils low in calcium may be acid and in these soils particularly those of heavy texture much aluminium is present in the soil solution. Aluminium is particularly toxic to roots and the root systems of plants growing in acid soil are often poor with short and stubby individual roots. Manganese is also present in large quantities in solution in acid soils (see Manganese).

Excess symptoms. Soils of high lime status rich in calcium have high pH values and may also contain bicarbonates; the effects of these on plant growth are complex. The main effects are induced iron deficiency and manganese deficiency. The use of sulphate of ammonia may be beneficial in these circumstances.

MAGNESIUM

Deficiency symptoms. Some symptoms of magnesium deficiency are strikingly similar to those produced by potassium deficiency and there is a risk of confusion, but the scorch symptoms due to potassium deficiency are often ashy in colour whilst those due to magnesium are more brown. About 10 per cent of the magnesium in a plant is present in chlorophyll and much of the remainder is present in enzymes affecting many activities in every living cell in a plant. In conditions of shortage the total amount of magnesium in any leaf remains constant but because of increase in size of the leaf the amount as a percentage of the dry matter decreases. Initially parts of the leaf become necrotic but eventually the whole leaf dies and falls off. Often by August the whole of an apple shoot will be bare except for a few young leaves at the tip. Shoot growth continues slowly but because of the defoliation shoots may look abnormally long. Where the deficiency is severe, shoot growth ceases and die-back symptoms appear. Apples from severely affected trees are mainly of cull grade and lack flavour and colour. They are woody in texture and tend to shrivel in store.

Leaf symptoms vary from species to species and with nitrogen status, and in apples in particular there is a wide variation between varieties. The symptoms form distinctive patterns and spread progressively from older to younger leaves. The main effects are chlorosis, and red and purple tinting followed by death of the tissue.

Typical leaf symptoms are:

Apple. Interveinal necrosis, brown blotching, beginning either near the midribs or the margins; the necrosis may be accompanied by chlorosis and/or purpling. The variety Edward VII shows symptoms very early in the season.

Cherry. Interveinal necrosis as for apple, but usually accompanied by high purple or red tinting.

Pear. Necrosis of the central areas of the leaves, leaving green margins but in some varieties the necrotic areas develop at the margins.

Plum. Interveinal central or marginal browning sometimes preceded by chlorosis; where the symptoms arise centrally, a narrow marginal area may remain green throughout.

Black currant. Intense purpling of the central area, leaving a clear green marginal band, but occasionally the purpling may begin at the margin; margins of affected leaves may become severely curled downwards.

Gooseberry. A broad red marginal band is developed, which extends towards the midrib and eventually fades to yellow; tissues along the midrib may remain green.

Excess symptoms. None have been observed in the field but high magnesium may increase the incidence of bitter pit.

SULPHUR

Symptoms of deficiency or excess have not been observed in the field in this country.

IRON

Deficiency symptoms. Because iron does not move easily in plants suffering from an induced iron deficiency the characteristic effect of this deficiency is the development of marked chlorosis of the youngest leaves, for example, at the tips of the shoots, the chlorosis decreasing progressively in the older leaves. The pattern of the chlorosis is also distinctive, especially in the earlier stages, the outstanding feature being that the veins form a green network in the chlorotic areas. As the chlorosis becomes more severe the vein pattern may be obscured and the whole leaf becomes ivory-white; areas along the margins of the chlorotic leaves die out forming brown patches. Shoots and branches bearing severely chlorotic foliage eventually die back and the whole plant may die. Apple and pear fruits may be chlorotic or highly flushed and may be low in sugars.

Excess symptoms. Have not been observed in this country.

MANGANESE

Deficiency symptoms. Chlorosis is a prominent symptom which begins near the leaf margins and progresses interveinally in V-shaped tongues towards the midrib. Unless the deficiency is severe, the midrib and the main lateral veins remain green. As manganese moves in the plant more easily than iron, leaves near the tips of shoots often remain green, and this effect is in contrast with the symptoms of iron deficiency. Where the deficiency is acute, shoot growth is reduced and die-back may occur. The deficiency has not affected the appearance of the fruit in instances observed in the field.

Excess symptoms. These can arise in fruit growing in strongly acid soils leading to measley bark conditions of apple trees and to extensive necrosis of leaves of other fruits; some varieties of strawberry exhibit a pronounced purpling along the veins.

BORON

Leaf symptoms of deficiency or excess have not been observed in this country though it has been suggested that low levels of boron in apples may be associated with poor storage qualities.

COPPER

Deficiency symptoms. These have been observed in a few restricted localities mainly on sandy soils, particularly those derived from the Upper Greensand, the Lower Greensand and the Bagshot sands. The terminal leaves of shoots are first affected, beginning in early summer. The young leaves become cupped, develop irregular necrotic areas and eventually wither and fall. This stage is followed by die-back of the shoots, which progresses rapidly from the tips, so that whole branches may become affected. The die-back of the shoots, with its rapid development from June onwards, has been named 'Summer Die-Back'. The bark becomes rough and deeply fissured and may peel off in places. As die-back progresses young shoots may develop from the basal portions of stems and these eventually die back also. In stone fruits, for example plums and cherries, there may be considerable distortion and gumming of shoots. Severely affected trees may die quickly. Pears seem more susceptible to deficiency of this element than other fruit crops.

Excess symptoms. Have not been observed on fruit trees in this country.

ZINC

Deficiency symptoms. Zinc deficiency has been observed, but rarely in this country. Leaves show some interveinal chlorosis and the margins tend to be wavy in outline. Internodes are short and shoots often die back. Where the deficiency is severe the terminal growths comprise rosettes of small, narrow leaves. These conditions have given rise to the terms 'rosette' and 'little leaf'.

Excess symptoms. In certain areas of the country, the soils naturally contain excessive amounts of zinc. The symptoms are similar to those of iron deficiency chlorosis. Strawberries suffer severely. In a few localities, soils contain excess zinc from industrial contamination and from applications of contaminated sewage sludge.

MOLYBDENUM

Symptoms of deficiency or excess have not been recorded in fruit crops in this country.

Variation between Plants

Plant roots differ in their ability to forage for nutrients. The strawberry is very efficient in extracting phosphorus from soil and it is possible to grow full crops of strawberries on soils where phosphate deficiency would reduce the yield of black currants.

Marked differences have been observed in the susceptibility of certain apple and pear rootstocks and scion varieties of apples to mineral deficiencies. Among the older series of Malling apple rootstocks, M.5 has been found outstandingly susceptible to potassium deficiency. M.1, M.2, M.4 and M.7 are also susceptible, while M.9 and some of the vigorous stocks are less so. Unworked rootstocks which have been found to be highly susceptible to magnesium deficiency are M.1, MM.106, M.4 and Crab C. Rootstocks M.9, MM.111, and M.8 are less susceptible and stocks M.16, M.2 and MM.104 are fairly resistant to deficiency of this nutrient. Stocks worked with Jonathan and Cox's Orange Pippin have confirmed the high susceptibility of M.1, M.4 and MM.106 rootstocks. Worked stocks of M.7, M.25, M.16 and MM.111 were less susceptible whilst worked stocks of M.2 and MM.104 were fairly resistant to magnesium deficiency. The cherry rootstock F.12/1 often shows this deficiency especially in the nursery. On copper deficient soils, apples worked on the vigorous rootstocks M.12 and M.16 have been more susceptible than those on M.2.

In comparative tests Cox's Orange Pippin, Beauty of Bath, Early Victoria, Grenadier, and Newton Wonder were found to be highly susceptible to potassium deficiency. Worcester Pearmain and Bramley's Seedling were more resistant. Cox's Orange Pippin, Miller's Seedling, Allington Pippin, Grenadier, Edward VII, Lord Derby and Lane's Prince Albert are more susceptible to magnesium deficiency than other varieties. Differences have also been noted in other fruits. For example, pears on most quince stocks are more susceptible to iron deficiency than on pear stock.

The raspberry Norfolk Giant is very susceptible to both iron and manganese deficiency and the gooseberry variety Careless readily shows symptoms of magnesium deficiency.

Soil, Leaf and Fruit Analyses

Soil Analysis. Soil analysis can give useful information provided representative samples of soil are taken from the area. Because of the long-term nature of fruit plantations it is advisable to sample at 0–6 in. and 6–12 in. depths so that a measure of the nutrient level of a part of the soil profile can be obtained. It is particularly important that soil samples should be taken when a new orchard or plantation is being considered so that the opportunity can be taken to correct any gross deficiency of lime or nutrients and, if necessary, of cultivating in the remedial dressings.

Soil analysis is the best way to detect acidity but the results of soil analysis can also serve as the basis for deciding future applications of phosphorus, potassium and magnesium. At the present time, soil analysis does not provide a sound basis on which to judge the nitrogen requirement of fruit crops.

Leaf Analysis. Analysis of leaves can be of assistance in diagnosing nutritional problems and also in assessing the nutritional status, particularly the nitrogen status. To be of value it is essential to select leaves from a specified part of the tree or plant and, as with soil sampling, the sample must be representative of the crop in question.

Fruit Analysis. While the analysis of apple leaves can give some guidance on the likely storage properties of an apple crop, better information can be obtained from the analysis of a sample of apples taken at picking time. The aim when manuring fruit should be to supply sufficient nutrients for good growth and adequate yield of quality fruit without giving an excess of any one nutrient. Excessive amounts are wasteful in two ways: the nutrient may upset the uptake of another nutrient and so reduce either cropping or quality and the nutrient may be lost by leaching. Treatments should be based on the results of an analysis. Details of all sampling techniques and of the interpretation of analytical results can be obtained from the local ADAS horticultural adviser.

Correcting Deficiencies

There are three methods a grower can use to overcome the effect of deficiencies:

1. Supply the element by—
   (a) application to the soil
   (b) spraying on to the plant
   (c) injecting into trees

2. Correct an imbalance by—
   (a) supplying the appropriate element (by any method outlined in 1 above)
   (b) withholding supplies of an element present in excess
3. Changing soil management by—
   (a) reducing depth of cultivation
   (b) use of herbicide strips

Application to Soil

It is often possible to avoid deficiencies by taking care when selecting the site for a plantation. As mentioned above, soil samples should always be taken from two depths before planting. Sometimes the surface sample may show that sufficient nutrient is present but a deficiency may be shown lower down in the soil profile. Roots will be actively absorbing from quite deep in the soil and it is possible to work fertilizers and lime some way down into the soil before planting. Strong acidity must be corrected before planting by a suitable liming programme. If the soil is also short of magnesium then a magnesian limestone should be used. As mentioned on page 24 it is very dangerous to over-lime.

The application of fertilizers containing nitrogen, potassium and phosphorus is described on pages 26–27.

Sprays

Most plant nutrients can be absorbed from a solution sprayed on to the leaves. Sprays containing major nutrients should not be required in a well managed orchard but they can be used as a first aid measure, for example, urea at 5 lb per 100 gal can be sprayed on apples at about fortnightly intervals from petal fall onwards to provide nitrogen. Trace element deficiencies can be conveniently corrected by routine spraying but there is some risk of scorching the foliage so spray strengths must be correct. It is sometimes possible to mix nutrients with insecticides or fungicides but the manufacturers should be consulted to make sure that the ingredients proposed are compatible.

Injections

This method is seldom used in practice except for the correction of iron deficiency. It is possible to inject trees with powdered salts but it is more satisfactory and convenient to use tablets or pills. Commercial products are available in two sizes $\frac{15}{32}$ in. in diameter containing 1 g of the salt and $\frac{7}{32}$ in. in diameter containing $\frac{1}{8}$ g of the salt. An Irwin type bit is used $\frac{1}{2}$ in. diameter for 1 g pills and $\frac{1}{4}$ in. diameter for $\frac{1}{8}$ g pills. Corks about $\frac{1}{8}$ in. thick and either $\frac{1}{2}$ in. or $\frac{1}{4}$ in. diameter are required for closing the holes.

Injections should be made into holes, which are staggered in a vertical direction, spaced horizontally about the stem at about 3 inch intervals.

The position of the holes should be marked. The holes should be drilled, with the Irwin type bit of appropriate diameter, horizontally into the tree stem to a depth sufficient to take tablet(s) and cork. The depth should be such that the outside face of the cork is slightly to the inside of the surface of

Good growth of *Bramley's Seedling* apple on Langley soil series in Kent

[*Photos: East Malling Research Station*]

Poor (flat-headed) growth of *Bramley's Seedling* apple on Fairbourne soil series in Kent (same age as tree above)

PLATE I

Worcester Pearmain apple on Dunnington Heath soil series—soil shallow

[*Photos: Long Ashton Research Station*]

Worcester Pearmain apple on same soil series, but soil deep

PLATE II

Poor growth in strawberries due to soil compaction

[*ADAS Plant Pathology National Slide Collection*]

Bitter pit in *Howgate Wonder*

PLATE III

Potassium deficiency in apple

Potassium deficiency in red currant

Red currant: chloride injury from excess of chloride-containing fertilizer

PLATE IV

Magnesium deficiency in apple

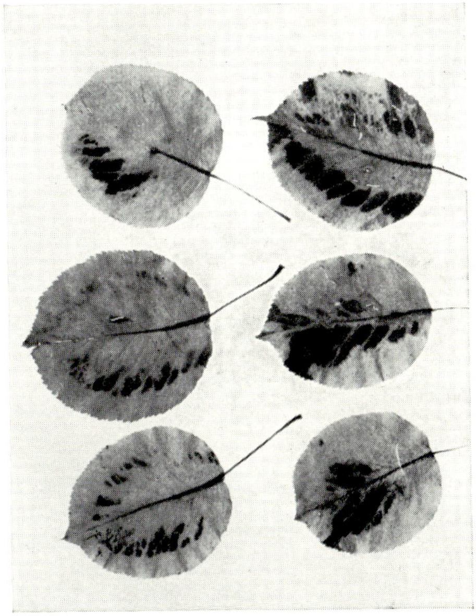

[*Photos: Long Ashton Research Station*]

Magnesium deficiency in pear

Magnesium deficiency in plum

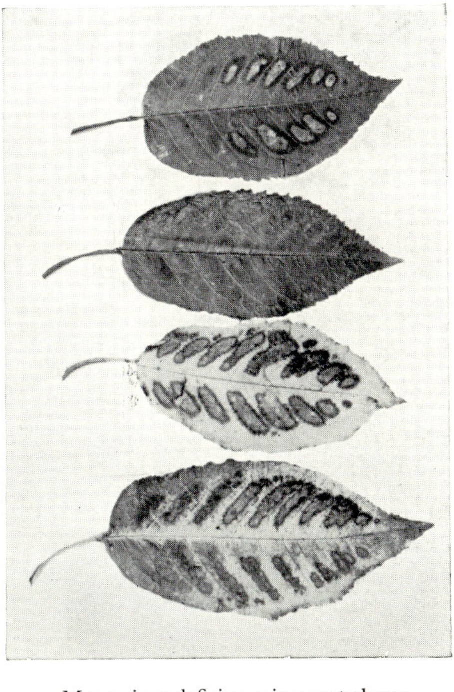

Magnesium deficiency in sweet cherry

PLATE V

[*Photo: Long Ashton Research Station*]

Iron deficiency in *Victoria* plum

Iron deficiency in apple. (Tip leaves chlorotic)

PLATE VI

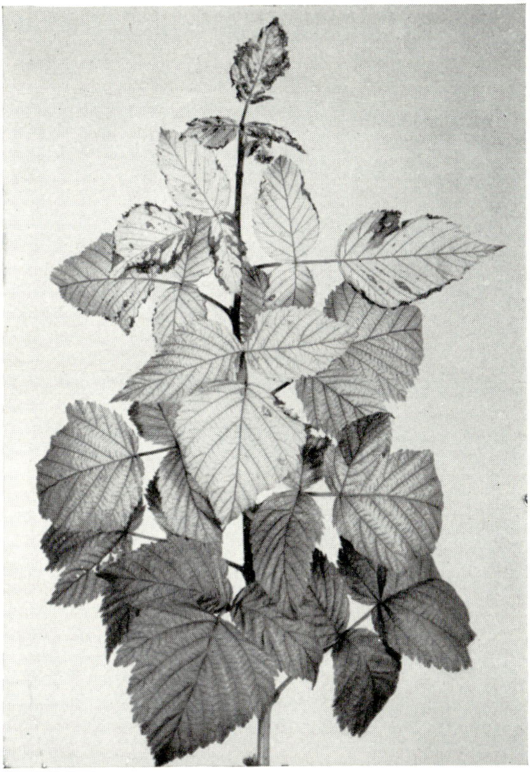

[*Photos: Long Ashton Research Station*]

Iron deficiency in raspberry. (Tip leaves chlorotic)

Manganese deficiency in apple. (Interveinal chlorosis)

PLATE VII

[*Photos: Long Ashton Research Station*]

Manganese deficiency in *Lloyd George* raspberry. (Interveinal chlorosis)

Manganese deficiency in sweet cherry

PLATE VIII

the wood of the stem. The hole should be sealed with grafting tape. In practice this system is only used for the control of iron and manganese deficiencies and the dosage for this is shown in the following table:

Suitable dosages of ferrous sulphate or manganese sulphate for injection as tablets

Diameter of stem inches	Total number of holes	Size of tablet gramme	Number of tablets per hole
1	1	$\frac{1}{8}$	1
$1\frac{1}{2}$	2	$\frac{1}{8}$	1
2	2	$\frac{1}{8}$	2
3–6	3–6	1	1
6–10	6–10	1	2
10–20	10–20	1	3

If a large number of trees have to be injected a special tool consisting of a thin walled steel tube of appropriate diameter fitted with an internal plunger will be useful.

INDIVIDUAL NUTRIENTS

Magnesium Deficiency

May be remedied by using magnesian limestone, where appropriate, Epsom salts or kieserite as soil applications but these take some time to become effective. Spraying treatments give more rapid effects and high volume spray is more effective than low volume. Four or five sprays of a 2 per cent solution of magnesium sulphate plus a non-ionic wetting agent every 14 days from petal fall onwards give the best control. The use of fertilizers providing nitrogen in the nitrate form alleviates the deficiency and there should be a carefully planned programme of manuring with potash fertilizers to prevent the build up of excessive levels of potassium in the soil.

Calcium deficiency

Useful partial control of bitter pit can be obtained with four to six applications of either calcium nitrate (10 lb per 100 gal) or calcium chloride (8 lb per 100 gal) with added wetter, applied at 200 gallons per acre. Worcester Pearmain and Merton Worcester spraying should begin in early June and for Cox and later maturing varieties in mid to late June. Spraying should continue at 21-day intervals as near as possible up to harvest. Young foliage is susceptible to scorch from calcium sprays and, to safeguard against this, the June sprays may be applied at half-strength at 10-day intervals. Provisional recommendations for low volume applications are 10–15 lb per acre (30–50 gal per acre respectively). Both at high and low volume it is essential to ensure complete spray coverage of the fruit.

Calcium chloride is as effective as the nitrate in controlling bitter pit but may occasionally cause more leaf scorch. The nitrate may slightly retard development of red colour and it may be better to substitute the chloride for the nitrate where maximum colour is required. Calcium nitrate can cause damage to the fruit of some varieties, notably Discovery, but the chloride does not cause such damage.

Low volume applications occasionally lead to more leaf scorch and, in such cases, more frequent applications at lower concentrations should be made. Both calcium salts are liable to damage apple foliage when temperatures are in excess of about 21°C (69–70°F) and if spraying cannot be deferred until evening, the concentration should be halved.

Calcium nitrate appears to be compatible at high volume with most wettable powders including captan and dinocap. Care must be taken to cream the wettable powder suspension before dilution to field strength and adequate agitation should be maintained, particularly for low volume application. Calcium nitrate is likely to be incompatible with magnesium sulphate, certain wettable sulphur formulations and liquid formulations of insecticides and acaricides.

Successful results without damage to the fruit have been obtained by the use of calcium nitrate to control bitter pit in Cox's Orange Pippin, Worcester Pearmain, Egremont Russet and Merton Worcester, all varieties subject to this disorder. On the other hand, fruits of Crispin (Mutsu) and Discovery have been seriously marked when sprayed with calcium. Damage has also been reported on several other varieties though in most cases the damage was very slight. Thus, before using calcium nitrate on varieties for which specific recommendations are at present lacking, preliminary tests should be made on a small area of orchard to ensure that there are no adverse effects.

Iron Deficiency

In the past it has not been possible to cure iron deficiency by ground applications of iron salts because soil conditions render the iron unavailable to plants. Formerly sprays were not very successful because of severe damage from scorching. However, of recent years, new compounds of iron known as chelates have become available, of which one of the most efficient is 'Sequestrene 138 Fe'. This can be used as a soil application at the rate of ½ oz per sq yd every 3 years or ⅙ oz per sq yd per year on an area covered by the spread of the branches. If applied in February there is usually no need to water it in but if applied later than this it should be well watered in.

Another iron chelate 'Sequestrene Fe EDTA' can be used as a foliar spray at ½–1 lb per 100 gal with a 'non-ionic' wetting agent. The first spray should be applied at petal fall and repeated three or four times at fortnightly intervals thereafter. Where the chlorosis is severe, the initial spray should be at the ½ lb rate to avoid scorch.

Iron sulphate can be injected into the stems of tree fruit but this method causes difficulties with plums because trees of this fruit often produce large amounts of gum after injection.

One of the effects of grassing down is to reduce iron deficiency.

Manganese Deficiency

Ground applications of manganese compounds are of no use to correct this deficiency because soil conditions make the nutrient unavailable to plants. Injection of manganese sulphate into the trunk or spraying with a solution of manganese sulphate are effective remedies. A concentration of 3 lb of manganese sulphate per 100 gal is used as a spray at petal fall and is usually sufficient to give control of the symptoms for the rest of the season.

Copper Deficiency

This can be cured with 28 lb (or up to 56 lb in severe cases) per acre of copper sulphate as a ground application or with a foliar spray of ½ lb of copper sulphate per 100 gal of water.

SUMMARY

Element	Fruit especially liable to deficiency	Suitable methods of control
Nitrogen*	Black currant, plum	5 lb urea per 100 gal of water sprayed at fortnightly intervals during summer.
Phosphorus*	Black currant	
Potassium*	Apple, gooseberry, raspberry, strawberry	
Magnesium*	Apple, gooseberry	20 lb Epsom salts per 100 gal of water 2 or 3 sprayings.
Calcium	Apple	10 lb calcium nitrate or 8 lb calcium chloride per 100 gal of water at 21-day intervals beginning mid-June.
Iron	Apple, pear, plum, raspberry, strawberry	Injection into stem. Iron chelate as a soil application.
		Iron chelate spray, ½ lb per 100 gal of water at fortnightly intervals.
Manganese	Sweet cherry, apple, raspberry, black currant	Injection into stem. 3 lb manganese sulphate per 100 gal of water, 2 or 3 sprayings.
Copper	Pear, plum, cherry	Soil application. ½ lb copper sulphate per 100 gal of water, one or 2 sprayings.

* See pages 59–66 for rates of fertilizer application.

Manuring in Relation to Cultural Practices

SOIL ANALYSIS AND MANURING PRIOR TO PLANTING

When a site is selected for fruit, a soil analysis giving the levels of phosphorus, potassium and magnesium, pH status and lime requirement should be obtained. Most sandy soils tend to be short of potassium and magnesium and require regular application of these nutrients. Many heavy soils are naturally well supplied with potassium and magnesium; some are rich in potassium. In nearly all soils the top soil is richer in phosphate than the subsoil due to an accumulation of applied phosphates and the difference tends to be most marked in heavy soils.

Since phosphorus, potassium, magnesium and lime move very slowly in the soil it is important to correct any major deficiency before planting so that it can be worked into the root zone.

The phosphorus requirements of most fruit crops are small but any marked deficiency should be corrected before planting by working in an application of about 60 units P_2O_5 per acre of phosphatic fertilizer.

Potassium and magnesium are important for all fruit crops and since magnesium deficiency can be induced or aggravated by an excess of potassium it is important to maintain a balance between them in the soil. The ratio of soil potassium level to soil magnesium (parts per million (ppm) K and Mg by ADAS methods) should not be much greater than about 2:1 to ensure that magnesium deficiency does not occur. When the subsoil is extremely deficient in potash it is advisable to plough down up to 5 cwt/acre (300 units K_2O/acre) of muriate of potash before planting.

Magnesium may be applied as one of the fertilizers described on page 33 or, where there is also need to supply lime, magnesian limestone may be used.

USE OF LIME

Slightly acid soil with a pH of about 6·0 is ideal for most fruit crops which will tolerate pH values of 5·5 or even below. Black currants are an exception; growth and cropping are adversely affected when the pH falls to about 5·5. For black currants acid soils should be limed before planting to give a pH of 6·5. For all other fruit crops a pH of 6·0 should be aimed for. Although pH levels considerably lower than these values can be tolerated it should be borne in mind that the soil test figure is an average for the orchard and there may be patches which are more acid. The average pH therefore should be maintained close to 6·0 (6·5 for black currants) to ensure that there are no patches acid enough to restrict growth.

Acid soils are often deficient in magnesium and when this is the case liming with magnesian limestone is advisable since it is a cheap form of magnesium as well as being an efficient liming material.

Although excess acidity should be corrected before planting, the lime applied is unlikely to last the life of the crop except for strawberries which are relatively short term. For these occasional liming will generally be necessary. When planting strongly acid soils with a pH below 5·0 it is advisable to plough down part of the lime requirement and cultivate the rest into the soil after ploughing.

Although correction of acidity is essential it is of the utmost importance to avoid over liming since this can lead to deficiencies of trace elements such as manganese and iron.

Under the grass/herbicide strip system of management it has been found that the herbicide strip tends to become acid rather more quickly than the grassed alleyways. So it is important to take soil samples for lime testing from the rows.

MANURING AFTER PLANTING

With newly planted bush and cane fruits a surface mulch of farmyard manure or other bulky organic material placed around the plants in late winter or early spring will generally help their establishment.

During the first year after planting fruit trees the aim is to reduce the check caused by transplanting and to develop a good root system. A mulch of farmyard manure or straw placed around the trees a little distance from the trunk will help to retain soil moisture and greatly aid re-establishment of the tree. If fertilizer applications have been made before planting, additional dressings are not usually needed in the first season. Provided the trees have been mulched or can be irrigated they should make satisfactory new growth and there is generally no need for nitrogenous fertilizers at this stage.

In the following years the aim with top fruit is to maintain vegetative growth so as to develop a satisfactory branch framework capable of producing and carrying heavy crops. Growth of the young tree is encouraged by eliminating weed competition, mulching where possible and by pruning designed to develop the branch system and at the same time encourage early cropping.

In the first few years after planting fruit trees, annual applications of fertilizers are required over the area explored by roots which generally extend considerably beyond the spread of the branches. Care should be taken to avoid excessive applications of fertilizers or concentrations near the trunk as they may be damaging. As the trees grow in size the area covered by the fertilizers should be extended so that within a few years the fertilizers can be broadcast over the whole orchard area. If there is any marked soil deficiency of potassium or magnesium which has not been corrected prior to planting, annual dressings should be spread over the whole area even in the early years so as to ensure adequate supplies for the extending root systems. When thinning the orchard is necessary, the disturbance of the root systems of the trees to be left and the resulting improved light conditions may lead to nitrogen deficiency for two or three years unless extra nitrogen is given.

In the early stages of tree growth, the trees require little nitrogenous fertilizer, provided there is a sufficient level of organic matter in the soil, but when the bearing stage is reached the requirements for nitrogenous fertilizers increase especially if an over-all grass sward is developed in the orchard. Apart from increasing the need for nitrogenous fertilizers an over-all grass sward can also result in greater availability of iron, potassium and phosphorus than the 'arable' system of clean cultivation which prevents root growth in about the top 3 in. of soil. When the soil around the trunk or a strip of soil in the row is kept bare by means of herbicides the effect of the sward on the nitrogen requirements is greatly diminished. The nitrogen requirement in the herbicide strip system is only a little above that under arable.

TIME AND METHOD OF APPLICATION OF FERTILIZERS TO DIFFERENT FRUITS

Apples

The question of the best time to apply the main mineral nutrients, especially nitrogen, to fruit crops has for long been the subject of much discussion. It has often been suggested that early spring application of nitrogenous fertilizers promotes vegetative growth, whereas mid-summer application at the time when fruit buds are developing promotes fruitfulness. It is known that an adequate supply of nitrogen is needed during the periods of flower bud differentiation and fruit set. With an over-all grass sward there is a risk of the trees suffering from a deficiency of nitrogen during the summer because of competition from the grass. In a long term experiment at Efford Experimental Horticulture Station dessert apples in grass responded best to nitrogen applied in August, but December applied nitrogen followed by February applied was almost as good. Nitrogen applied in April was relatively ineffective, no doubt because the grass used most of the nitrogen, while an application in June reduced the yield considerably, even below that obtained from trees given no nitrogen at all. The adverse effect of nitrogen applied in June may be due to enhanced competition for moisture by the sward at a time when grass growth is normally very vigorous.

Even though some experimental evidence suggests August application is a little better than winter applications the most sensible time to apply nitrogen would appear to be in late January or early February and this practice gives satisfactory results. At this time the grass is dormant and there is normally sufficient rain to wash the nitrogen well into the soil before grass growth commences. Application in August carries the risk of mechanical damage to the fruit and autumn applied nitrogen may be lost by leaching in wet winters. In arable systems or with the grass and herbicide strip system there is less chance of a severe nitrogen shortage in summer. Application in mid to late winter of the smaller amounts required is quite satisfactory.

Dressings of phosphorus and potassium may be applied any time during the winter or early spring whichever is more convenient. They act relatively slowly so that it is probably best for applications to be made while it is still likely that they will be washed into the soil surface by rain.

When magnesium is applied as magnesian limestone or where lime is to be applied, the applications should be made in winter. Soil dressings of magnesian fertilizers may also be applied then or with the other fertilizers in late winter or early spring. Magnesium sulphate (Epsom salts) used as a foliar spray is best applied as three or more sprays at two-weekly intervals, beginning at petal fall.

When soil levels of phosphorus, potassium and magnesium have been built up to the optimum and applications are for maintenance only, it is no longer necessary to apply these nutrients every year. Applications every two or three years will suffice for phosphorus and potassium and the interval may be even longer for magnesium.

The best time to apply bulky organic manures usually depends on convenience or any special purpose for which they may be used. In the normal way they may be applied during the winter but where used specifically as a mulch may be put on in early spring. Straw mulches are best applied towards the end of the winter when there is still a likelihood of rain which can help to settle the straw and reduce the risk of fire.

Pears, Plums and Cherries

Fertilizers for pears, plums and cherries, may be applied in the same way as for apples. There is at present no evidence of any effect of varying the time of application of fertilizers for these crops.

Soft Fruits

With soft fruits phosphatic and potassic fertilizers may be applied any time during the winter or early spring. Since soft fruits are not usually grown in grass there is not the same need to apply nitrogenous fertilizers so early as in grassed orchards, and February or early March applications are suitable.

There is no definite evidence of the value with soft fruits of applying nitrogen or other fertilizers at other times of the year as, for instance, after the crop is picked. With strawberries, in some instances where nitrogenous fertilizers have been applied when the land has been cleaned up after picking, vegetative growth has been encouraged with a reduction in cropping in the following year. In other experiments with strawberries there has been no response to nitrogenous fertilizers when applied either in the spring or after picking.

EFFECT OF PRUNING

Since pruning influences vegetative growth, it is advisable to consider pruning and manuring programmes, especially the use of nitrogen, in relation to one another. As a general rule severe branch thinning and shoot tipping in a mature fruit tree will tend to promote growth, in a manner similar to heavy dressings of nitrogenous fertilizers, and the combination of hard pruning and manuring may result in over-vegetative plants. Where the pruning of mature apple and pear trees has been neglected over a long period, very severe cutting out and shortening of old spur systems may result in a temporary depletion of food reserves within the tree. It is better, therefore, to spread the pruning of such trees over several years.

More generous manuring may be necessary with the regulated method of pruning apple trees in order to ensure sufficient new growth. Often a stimulus to growth may be better obtained by pruning treatments than by heavier applications of nitrogenous fertilizers. For example, with young trees which are tending to produce excessive quantities of fruit buds and insufficient shoot growth, removal of some or all of the fruit buds at pruning time or removal of the flower trusses later will promote new shoots and wood growth. In such cases removal of the fruit buds or flowers alone, with little pruning of the shoots, is often best and sufficient to encourage vegetative growth. It should be remembered that hard pruning not only reduces the size of the branch system but also has a restricting effect on root growth.

Black currants and other bush fruits usually respond to hard pruning by making new shoot growth though, if soil moisture and availability of nutrients are limiting growth, severe pruning alone is unlikely to give satisfactory results. Moderate pruning coupled with adequate nutrition and water will generally give best yields.

With cane fruits such as raspberries and loganberries limitation of the amount of new cane retained for cropping will usually prove best and ensure better sized berries than where all new growths are left to fruit resulting in severe competition between the canes for water and nutrients. Excessive competition in this way may also reduce growth of new cane for the next season's cropping.

INTERCROPPING

Although at one time common, mixed plantings of fruit crops are now comparatively rare owing to the difficulties they create in management due to different pesticide and nutrient requirements. However many growers who are starting a fruit growing enterprise with limited land and capital available, find the years of waiting before harvesting their first tree fruit crop a difficult period, and welcome any income that can be derived from soft fruit crops. If soft fruit intercrops are used, the nutritional needs of the crops should not differ too greatly from those of the trees, for example with dessert apples, strawberries can be grown successfully with the least harm to the apples, while plums and black currants are fairly well suited because of their needs for a similar level of manuring, particularly for nitrogen. Strawberries may be grown as an intercrop with young cherries. Where intercrops are grown, the needs of the permanent trees should come first and intercropping must be discontinued before it interferes with the growing of the trees.

PEST AND DISEASE CONTROL

Although the view is sometimes held that diseases occur only when nutrition is faulty, the general experience in fruit growing is that, whether the plants are manured with organic manures or with fertilizers, well-nourished trees in the right condition for bearing heavy crops of high quality fruits are always subject to attacks of pests and diseases. Fruit plants over generously supplied with nitrogen are, however, especially liable to insect pest and fungus disease attacks.

Gloeosporium spp, a fungus causing storage rots in apples, is often worse where heavy applications of nitrogenous fertilizers have been used. In the same way vegetative vigour in apples induced by these fertilizers may aggravate attacks of apple canker caused by the fungus *Nectria galligena*.

At present there is no known method of manuring fruit plants that renders them sufficiently immune from pests and disease to dispense with pest and disease control measures. For this reason adequate manuring must always be accompanied by efficient pest and disease control practices.

Potash manuring is sometimes regarded as a means of rendering plants resistant to disease, and although this may be true for some diseases of fruit plants, it cannot be accepted as applying generally. Thus, black currant leaf spot (*Pseudopeziza ribis*) was found at Long Ashton Research Station to occur only to a negligible extent on acutely potash-deficient plants and to be severe on comparable plants to which potash fertilizer was given annually. Moreover, a deficiency of either nitrogen or phosphorus increased the severity of the attacks, but where nitrogen, phosphorus and potassium were all applied, attacks were still severe indicating that high potassium was the dominant factor determining susceptibility to attack.

Since spraying of commercial crops is essential, the relationship between spray damage and manuring is of importance, and the indication is clear, both from experiments and general experience in the field that a deficiency of any one of the essential mineral nutrient elements increases liability to damage from spraying. On experimental plants in sand culture, where tests were restricted to nitrogen, potassium, phosphorus, calcium and magnesium, deficiencies of potassium and phosphorus were associated with the most severe damage, but lack of nitrogen, magnesium and calcium all produced increases in damage. In the field, severe damage has been noted frequently in plants deficient in potassium or magnesium, and sometimes also in nitrogen.

EFFECT OF SOIL MOISTURE

The amount of water available to plants can have a profound effect on their growth and where water is a limiting factor, growth will be restricted. With fruit crops in which the crop is usually borne on growth formed during the previous year, the lack of water in a drought year may affect not only the crop in that year but also in the year following.

Under conditions of low soil moisture nutrients may be unavailable for use by the crop even from soils well supplied with them. Frequently under dry conditions heavy applications of nitrogenous fertilizers produce little or no response in the crop plant although with sufficient moisture growth would be vigorous. In practice it has been found that adequate supplies of water applied as irrigation may result in a lower need by the plant for nitrogenous fertilizers. A mulch of straw or other organic material to conserve soil moisture can have a similar effect.

Limitation in moisture supplies can markedly affect the cropping of bush and cane fruits and in particular black currants and raspberries benefit from irrigation or mulching even in years of moderate summer rainfall. The amount of soil water reserves available for use by the crop will depend largely on the depth and nature of the soil, in addition to the rainfall, so that the need for irrigation or mulching must be related to local conditions. With the strawberry crop in particular, response to irrigation appears to be dependent on the nature of the soil.

MANURING FOR YIELD AND QUALITY

In fruit growing the quality of the produce is of great importance, and it is useless to produce high yields of fruits that are unsuitable for particular market purposes. Quality may comprise a number of aspects, including colour, uniformity of size, freedom from blemishes, eating quality, storage properties and suitability for transporting and handling under marketing conditions.

Although nitrogen fertilizers are usually required to obtain maximum yields, over-generous use often produces fruits of relatively low quality, oversized, poorly coloured, soft and easily bruised, and with poor storage properties. The quality of apples, cane fruits and strawberries, and to a lesser extent pears and cherries, is particularly liable to damage by over manuring in this way. Black currants are less affected than any of the other fruits, and with this fruit, high yields and good market quality are closely related.

Soil Management

IMPORTANCE OF ORGANIC MATTER

The organic matter of soils is mainly confined to the top few inches of top soil and has several important functions. It is a mixture of fresh plant and animal remains and 'humus' which is the partly rotted dark brown material. Humus gives top soils their characteristic darker colour than the subsoil.

The effects of organic matter in the soil are very complex and are mainly due to the humus portion. Humus has important physical properties: it can absorb many times its weight of water and in addition has powerful cohesive properties. A good humus content improves the waterholding capacity of sandy soils which tend to be drought susceptible while the cohesive properties of humus are of value in fine sands and silts which tend to be unstable and 'slake' or collapse when wet.

One of the most important effects of organic matter in the soil is the maintaining or building up of a good crumb structure. Generally speaking, this means the aggregation of the finer particles of soil into crumbs which are of a size to give suitable pore spaces for the retention of moisture, but allow the draining away of excess water, the free movement of air and the easy penetration of roots. A good crumb structure remains stable under repeated wetting and drying and resists crushing by farm implements and tractors. Good soil structure and aeration allow the penetration of roots through the soil, which is essential for access to sufficient water.

Continued arable cultivation with no addition of any form of organic matter can in some soils, over a period of years, lead to severe deterioration in structure and fertility and on sloping sites may result in soil erosion. Failure to maintain a satisfactory level of organic matter is particularly serious with silty and sandy clay loam soils, where compaction of the soil surface is often a trouble, and also with gravel soils. In heavy soils the level of organic matter is usually much higher and cropping in the absence of organic manures may be carried out for longer periods without the same harmful effects provided soil drainage is good.

Organic matter is the main form in which nitrogen can be stored over a long period in the soil and its gradual decomposition by micro-organisms results in the liberation of ammonium nitrogen and the production of nitrates available for the growth of crops. In addition other plant nutrients such as phosphorus, potassium, calcium, magnesium and micro-nutrients may be present and become available to plants as decomposition proceeds in the soil. Humus also acts as a temporary storehouse for nutrient elements that are applied to the soil or released by decomposition. Most of the living organisms in the soil require organic matter, and soils which are active biologically are usually of high fertility.

The important part played by organic matter in the maintenance of soil structure is a major factor in the management of the soil in soft fruit plantations and orchards which are clean cultivated. When weed control is achieved entirely by herbicides the lack of disturbance by cultivations allows the soil to build up and maintain a stable porous structure and it is unnecessary to add organic manures to maintain soil organic matter levels. It is essential, however, to start a new plantation or orchard with good soil structure and a high level of organic matter helps to ensure this.

Grass leys are of particular value in improving the structure of soils low in organic matter and can be beneficial on sites intended for planting with fruit. The longer the ley can be left before ploughing up the greater its contribution to soil improvement. Prior to the planting of soft fruits, whenever possible, it is advisable to apply up to 30 tons or more per acre of farmyard manure. Such a dressing will often supply the major needs of a strawberry plantation throughout its three cropping years and will provide a good start for other soft fruits.

THE USE OF MULCHES

Apart from the incorporation of farmyard manure, compost and other organic materials in the soil before planting or during the life of the crop, these materials are often of great value when used as surface mulches. The mulch is generally limited either to the area of ground immediately around the plant or to the rows of plants, but is sometimes spread over the whole area as for instance in the case of an overall straw mulch of a black currant plantation.

Mulches have several important functions and can have a profound effect on the growth of the crop and its manurial requirements.

The major effect of a mulch is on the superficial soil beneath in retaining moisture. The rate of diffusion of water vapour through a loose mulch is much slower than the evaporation rate from the bare soil surface. Thus a wet bare soil may lose ½ in. of water in 3–5 days, whereas under similar conditions, a mulched soil will take about 6 weeks to lose this amount.

The beneficial effects of a mulch have been found in the case of raspberries and black currants to result in an increase in cane, bush growth and cropping. A mulch can also be of great benefit around newly planted trees as at this stage a satisfactory water supply is essential for the young trees to develop a new root system rapidly.

Apart from its effect in retaining moisture when applied to the surface of a soil already wet, a mulch usually allows better penetration of rain or irrigation water than a dry, bare soil and reduces waste of water by run-off. The use of uncomposted sawdust as a mulch may, however, as a result of its compact nature, actually prevent penetration of rain.

Mulches are of great value on silty or other soils subject to surface compaction. On such soils even a very light mulch will break the force of the rain or irrigation water droplets, so preventing compaction and by protecting the soil from exposure to alternate rain and sun, reduce the chance of cracking.

On badly drained soils and in very wet weather a mulch may have the effect of retaining too much water so that the soil becomes waterlogged and aeration of the soil is seriously reduced. Mulches should not, therefore, be used indiscriminately, or over the whole surface on soils known to suffer from impeded drainage, and their application after prolonged wet periods should be delayed until the soil has dried sufficiently. Raspberries are especially sensitive to the effects of impeded drainage and should be mulched with care under such conditions.

Effect of Mulches on Soil and Air Temperatures

Under organic mulches such as fairly loose straw or sawdust, soil temperatures are rather lower during the day and slightly higher at night than under bare soil. The daily temperature range is much narrower as a result of the insulating effect of the mulch. This may result in cooler soil conditions early in the season before the ground has warmed up and a slight retardation in growth of the mulched plants. In straw mulched raspberries there is delay in the growth of the new cane though ripening of the berries above the mulch is earlier than above the bare soil. The insulating effect of a mulch can also increase the frost risk to crops planted on frost susceptible sites. By slowing down the rate of the passage of heat from the soil to the air during radiation frosts, the air temperature at 2 ft above a straw mulch may be up to one degree C or 2 degrees F lower than that at the same height over bare soil. The temperature above a grass sward, which can also have an insulating effect if left uncut, is usually slightly lower than where the grass is mown short.

Effect of Mulches on Soil Nutrients

In addition to the effect of a mulch in conserving moisture so that the crop benefits from an increased water supply and can make better use of soil nutrients, the mulch can also have a direct effect on the availability of nitrogen. During dry periods, organic matter in the moist soil under a mulch can continue to decompose and nitrification continues longer than in a bare soil though in rainy weather increased penetration of rain into a mulched soil could lead to more leaching of nitrates from the surface. The moisture conditions under a mulch favour the development of soil micro-organisms which decompose the organic matter and help to maintain and improve the soil structure. Earthworms are encouraged, and by their burrows increase water and air penetration into the soil.

Another feature frequently observed with mulched soils is an increase in the amount of available potassium as the mulching material is usually rich in potassium. This increase has been recorded down to a depth of 8 in. after 2 years under a straw mulch. Similarly in apple orchards where the grass is mown and allowed to rot where it lies, an increase in the available potassium and phosphate is also commonly found. This has been reflected in an increased uptake of potassium and phosphate by the trees compared with those kept under clean cultivation or sown to annual cover crops.

The effect of mulches in keeping the surface soil moist and increasing the supply of available nutrients creates a very favourable environment for root growth and with the well aerated conditions normally present near the surface under an uncompacted mulch, surface rooting is encouraged. Excavations of black currant bushes grown under a mulch have shown development of a mat of surface roots whereas outside the mulched area the roots penetrate much more deeply into the soil. The use of a mulch usually eliminates the need for cultivations so that roots growing near the surface remain undisturbed.

Straw Mulches

Dressings of unrotted straw are often used for mulching both fruit trees and bush and cane fruits. Initially about 8 tons are required to cover one acre, subsequently 4 tons may be sufficient to maintain a soil cover. Straw by itself has very little nutritional value—one ton contains about 11 lb of nitrogen (N), 4 lb of phosphoric acid (P_2O_5) and 22 lb of potash (K_2O). Because it is rich in carbohydrates which require nitrogen for the rotting process, no nitrogen becomes available after the straw has rotted.

The loose nature of a fresh straw mulch increases its insulating effect and increases the risk of frost damage to the crop more than a compact mulch of well rotted farmyard manure.

THE EFFECT OF HERBICIDES ON SOIL STRUCTURE

The use of soil-applied persistent residual herbicides on the surface of clean ground to control germinating seedlings can maintain it in a weed free condition for the greater part of the year, so making cultivations unnecessary. The absence of any weed cover and the bare condition of the soil will tend to deplete the soil of organic matter but provided the soil is not compacted by wheelings and is not subject to waterlogging a stable and sufficiently porous structure develops even on inherently unstable silty soils.

GRASS SWARDS

The excellent soil structure under grassland is due to the extensive penetration of roots throughout the soil and the build up of organic matter from dead and decaying roots. The roots alone of the grass may contribute over one ton of dry matter per acre per annum apart from the material contributed from the top growth and when the grass sward is cut and the mown grass allowed to lie where it falls, as is the normal practice in grass orchards, the rotting of the grass makes a further contribution to the soil. Frequent mowing restricts the depth of rooting of the grasses, reduces competition between the grass sward and the trees for soil moisture, and limits competition for nutrients, in particular nitrogen, to the upper layers of the soil.

Effect of the Grass Sward on the Nutrition of Fruit Trees

When trees have to meet competition from a grass sward at a very early age, they may suffer a severe check and fail to grow away satisfactorily for a number of years. This can be avoided by maintaining weed-free strips in the tree rows or squares around the trees through the use of herbicides.

Mulching of the trees at time of grassing and for a year or so subsequently will also help them to meet the competition and maintain satisfactory growth. Normally orchards are planted on arable land or in cultivated strips in grassed fields. It is not usual to grass down until the trees are 5–7 years old, and then only the inter-row area. Additional applications of nitrogenous fertilizers during and following the year of grassing may be necessary to compensate for competition by the grass for nitrogen.

Once the sward is established it can be effective in reducing the severity of mineral deficiency symptoms. The level of available potash is increased while symptoms of iron deficiency are frequently less acute following grassing. An immediate effect of the initial grass competition with the trees is to reduce the level of nitrogen in the leaves and subsequently the level of both potash and phosphate is increased. The long term effect of a grass sward on apples is to improve the colour of the fruit and to reduce pre-harvest drop, so increasing the amount of picked fruit. The grass tends to delay the time of blossoming.

Other benefits of the grass are to provide a better working surface for fruit farming operations throughout the year and the maintenance of a better soil structure with less risk of surface compaction than under clean cultivation.

Effect of the Grass Sward and Non-cultivation on Root Growth

Where grass is allowed to grow in an orchard previously kept under clean cultivation or a change to non-cultivation with herbicides is made a change takes place in the root system of the trees. Free from disturbance by cultivations, some roots grow into the surface soil and develop a fine fibrous system in the upper soil layers which are normally the richest in nutrients. Under cultivation, a fine layer of roots usually grows just below the depth of the cultivations, any upward growing roots being pruned off by the implements.

Where an old established sward in an orchard is ploughed up, considerable damage is likely to be caused to the tree roots and there may be a check to the growth of the trees. Subsequent rotting of the old turf will, however, generally be followed by an increase in growth. Breaking up of established swards in this way may be of value in encouraging new growth where this is lacking, but it is best to use a rotary cultivator and confine the cultivation to the top few inches so as to minimize root damage. Where old turf is broken up clean cultivations may be carried out for a few years before a sward is again established or the grass may be allowed to grow again the same year it is broken up so as to reduce the likelihood of a new check to growth when the sward is re-established.

Types of Sward

When contemplating grassing down an orchard, local conditions should be studied carefully. Under conditions of low summer rainfall and light soil the checking effect of the grass due to competition for water may be too severe.

A mixture of perennial ryegrass with the addition of wild white clover has frequently been used for the grassing down of orchards, but it is often found that the perennial ryegrass causes a severe initial check to growth and its use is best confined to vigorous trees growing in deep soil. S.50 Timothy is less competitive in the early stages and has the advantage of being more easily managed; it is usual to sow wild clover with the timothy.

In many circumstances a sward developed from natural grasses proves quite successful. Generally at the outset the sward consists mainly of annual meadow grass (*Poa annua*) with annual broad-leaved weeds. They do not compete so seriously with the trees during drought periods as, for example, a sward of perennial ryegrass and the initial check to growth commonly seen in the year following grassing down is lessened. A light sowing of wild white clover may be used to supplement the natural grasses and as a leguminous plant, capable itself of fixing nitrogen, may add to the nitrogen supply available to the trees. Wild white clover is not usually very persistent and often disappears after a few years especially where heavy dressings of nitrogen fertilizer are used. In general the natural sward of annual meadow grass and weeds is gradually invaded by other indigenous grasses of which the most common is creeping bent (*Agrostis stolonifera*). Even when a specific grass mixture is sown at the outset, creeping bent frequently becomes dominant in the sward after a period of 5 to 7 years, but the ultimate nature of the sward will depend on local conditions.

USE OF LIVESTOCK IN ORCHARDS

In orchards of standard trees of cherries in Kent, the grazing of the grass by sheep has been a long established tradition and sheep are also often allowed to graze for part of the year in orchards of standard culinary apples. Cattle are used for grazing cider apple orchards in the West Country, especially in Hereford.

Except in these instances the use of stock in orchards is not recommended since it is apt to interfere with the general management and spraying of the trees quite apart from possible damage to the trees by stock.

Where livestock are employed their manure will contribute to the nutrition of the trees and will reduce the need for dressings of fertilizers. It may, however, be necessary to apply additional phosphates to encourage a satisfactory sward for grazing.

Fertilizers, Manures and Liming Materials

FERTILIZERS, manures and liming materials commonly used in fruit growing are described in the following sections but no attempt has been made to deal with all the materials which might be used occasionally.

INORGANIC NITROGENOUS FERTILIZERS

These contain the plant nutrient nitrogen (N) in a relatively cheap form. They are simple salts, readily soluble in water and quickly available to plants. Under Part IV of the Agriculture Act 1970, suppliers of such fertilizers are required to state the amount of nitrogen they contain. The principal fertilizers are as follows:

Ammonia (20 or 82 per cent N). At normal air temperatures and pressure, ammonia is a gas but it is readily soluble in water giving aqueous ammonia (20 per cent N) and it can also be liquefied under pressure to give anhydrous ammonia (82 per cent N). In either form it must be injected into the soil to avoid loss of ammonia to the atmosphere. Because of the highly corrosive nature and of the pressures involved with anhydrous ammonia, special injection equipment is needed. In established orchards the passage of the tines of the injection equipment through the soil can cause extensive damage to the root systems of perennial fruit crops and the use of ammonia fertilizers cannot be recommended.

On being applied to the soil, the ammonia is adsorbed by clay and organic matter and, to this extent, is preserved from immediate loss during wet weather. However, when the soil is warm enough, the ammonia is quickly changed to nitrate which is easily leached but this conversion ceases, or is very slow, below a soil temperature of about 5·5°C (42°F).

Sulphate of Amonia (21 per cent N). This is a white or greyish crystalline powder, the colour depending on the source of supply. Most sulphate of ammonia is manufactured synthetically from nitrogen in the air and some is a gas works by-product. Sulphate of ammonia (and other ammonium salts) is not quite so quick acting as those fertilizers which contain nitrogen in the form of nitrate because ammonia has to be converted by soil bacteria into nitrate before most plants can make use of it. As is mentioned above, the conversion of ammonia to nitrate depends on the soil temperature. Sulphate of ammonia tends to make the soil acid, 100 cwt of sulphate of ammonia causing the loss of an equivalent weight of ground limestone or chalk (calcium carbonate, $CaCO_3$). The acidifying effect of sulphate of ammonia is an advantage for fruit growing soils containing an excess of lime. On other soils, however, repeated heavy dressings of sulphate of ammonia may make the soil too acid for satisfactory growth of crops such as black currants and young apple trees.

Ammonium Nitrate and Ammonium Nitrate/Lime Fertilizers (21–35 per cent N). Ammonium nitrate contains half its nitrogen in the form of ammonia and the remaining half as nitrate. It therefore combines the properties of both these constituents. It has some nitrate for immediate action and the ammonia, while readily available, mainly takes effect somewhat later. The granulation of ammonium nitrate with calcium carbonate results in the production of an easily handled material known as ammonium nitrate/lime mixture (21–26 per cent N). The calcium carbonate present largely balances the acidifying effect on the soil of the ammoniacal part of these fertilizers. The granular form makes them easy to handle and spread but they become pasty if they are left exposed to the air for long periods.

In a further group of fertilizers, the ammonium nitrate is protected in a 'prilled' form to give a free-flowing material with a nitrogen content of about 34·5 per cent N.

Ammonium nitrate is also used as the source of nitrogen in some compound fertilizers.

Urea (46 per cent N). This is a highly concentrated nitrogen compound which is manufactured synthetically. Although an organic type, it is included here because it is similar in its general reactions to the inorganic nitrogenous fertilizers. It is available in both granular and powdered forms and, because it is highly soluble, is sometimes used in solution as a foliar spray on fruit crops. Urea may be less efficient on soils containing free calcium carbonate because of loss of ammonia if the urea remains on the surface of the soil.

Recently a form of urea fertilizer in which the urea has been coated with sulphur has been marketed as a slow release fertilizer.

Ammonium Phosphates. See later section on Phosphatic Fertilizers.

Nitrate of Soda (16 per cent N). This fertilizer is obtained by the purification of a natural salt deposit in Chile and is now marketed in a granular form. All the nitrogen is present as a soluble nitrate and as such is immediately available to crops as soon as it has dissolved in the soil moisture.

On soils of unstable structure (for example some soils containing a high percentage of silt and fine sand) and on clay soils, heavy and repeated dressings of nitrate of soda may spoil the tilth of the surface soil. It is only occasionally and under special circumstances that this effect is likely to be important in practice. Nitrate of soda has no tendency to make the soil acid.

Potash Nitrate (15 per cent N, 10–15 per cent K_2O). This is a mixed nitrate of sodium and potassium. A purer form of potassium nitrate can also be obtained and contains 13 per cent N and 45 per cent K_2O.

Organic Nitrogenous Fertilizers

Organic materials, such as meat-and-bonemeal, fish manures and manures derived from oil cakes and meals are very occasionally used as sources of nitrogen. They also supply some phosphorus and potassium. The total availability of nitrogen in these fertilizers is likely to be less than in inorganic fertilizers such as sulphate of ammonia. Moreover the conversion to nitrate in the soil is spread over a longer period although here much depends on the fineness of grinding of the organic fertilizers and other factors. The purchased organic fertilizers are usually fairly concentrated and are applied at relatively low rates per acre, hence the amount of organic matter added to the soil is negligible. They are mainly to be valued, therefore, for their plant food content but are expensive in this respect compared with inorganic fertilizers. They are usually applied in December and January, which is somewhat earlier than the inorganic nitrogenous fertilizers. Under the Agriculture Act 1970 the nitrogen (N) content of certain organic fertilizers has to be declared.

In addition, various kinds of animal wastes are used as fertilizers. For example poultry manure is sometimes used. It is variable in composition and in the fresh state may contain 70 per cent of water but only about 8 per cent if kiln-dried. The wet material is difficult to handle and apply evenly. The average composition of the fresh material is about 1·5 per cent nitrogen, 1–2 per cent phosphoric acid and 0·6 per cent potash while the kiln dried product contains about 3·5 per cent nitrogen, 3 per cent phosphoric acid and 1·5 per cent potash. Ammonia is easily lost in storage and some is lost on kiln-drying. The dried material is easier to handle and distribute and is more uniform in composition. In practice, fresh and dried poultry manures are good sources of nitrogen and phosphate and can be used successfully on crops such as plums and black currants which have a high requirement for nitrogen.

See also deep litter poultry manure in the Bulky Organic Manures section.

Feathers have a nitrogen content of about 8 per cent and are very occasionally used. Small feathers are easily decomposed in soil but larger ones are more persistent and less active.

Phosphatic Fertilizers

These fertilizers can be divided into two broad classes, those in which the phosphates are soluble in water and those in which they are insoluble. In the latter class, certain fertilizers contain phosphates which are insoluble in water but are soluble in weak acids such as citric acid. Water-soluble phosphates are most rapidly available to plants, followed by citric-soluble phosphates while those insoluble in both water and in citric acid are least rapidly available. When applied to the soil, however, water-soluble phosphates revert to less available forms and this reversion takes place most quickly in strongly acid soils. Phosphatic fertilizers can be applied to fruit at any convenient time during the year, generally this would be during the winter months.

Under the Agriculture Act Part IV 1970, a statement is generally required of the amount of total phosphate present (expressed as phosphoric acid, P_2O_5) and of the amounts of P_2O_5 soluble in water and in citric acid.

Superphosphate (18–19 per cent water-soluble P_2O_5). This widely used phosphatic fertilizer is available in granular form. In its manufacture, finely ground phosphatic rock is mixed with sulphuric acid to produce a mixture of water-soluble calcium phosphate (25–30 per cent) and gypsum (calcium sulphate, 60 per cent).

Triple Superphosphate (47 per cent water-soluble P_2O_5). This material in which the phosphate is water-soluble contains very little gypsum. It is obtainable mainly in the granular form. It is made by using orthophosphoric acid instead of sulphuric acid for converting the phosphate in the rock into a soluble form. As its name implies, it contains about three times as much phosphate as ordinary superphosphate, one cwt of triple superphosphate providing the same amount of water-soluble phosphoric acid (P_2O_5) as 2½ cwt of ordinary superphosphate.

The phosphates in ordinary and in triple superphosphate being soluble in water are rapidly available to plants when applied to the soil. These materials are widely used in the manufacture of compound fertilizers.

Ammonium Phosphates. Mono and di-ammonium phosphates are occasionally used in the pure state as fertilizers but, more commonly, they are constituents of many concentrated compound fertilizers. Both the phosphate and the nitrogen are soluble in water.

Ground Mineral Phosphates (40 per cent P_2O_5). Although the phosphorus in ground rock phosphates is not water-soluble, the rate of availability in the softer North African rocks, ground to pass a 100-mesh sieve, should be sufficient for fruit provided that the soil is not very deficient. The availability is highest in acid soil.

Basic Slag (18 per cent P_2O_5). The phosphorus in basic slag is assessed as 'citric acid soluble'. Although only slowly available under some conditions, the rate of availability should be adequate for fruit.

Meat-and-Bonemeal. This material, mentioned earlier under organic nitrogenous fertilizers, contains slowly available insoluble phosphates.

Potash Fertilizers

Like phosphatic fertilizers potash fertilizers can be applied at any convenient time during the year excepting that if broadcast during the summer there would be a danger of scorching the leaves of soft fruit crops.

Under the Agriculture Act 1970, the potassium content of potassium salts and mixtures sold as fertilizers must be declared in terms of the amount of potash (K_2O).

Muriate of Potash (60 per cent K_2O). This fertilizer which is almost pure potassium chloride is obtained in two forms, an ordinary salt and a free-flowing type known as 'pink flotation potash', specially adapted for drill application.

Muriate of potash is suitable for tree fruits, black currants and, at moderate rates of dressing (not exceeding 2 cwt/acre) applied during the winter months, for strawberries. Muriate of potash is unsuitable for raspberries, red currants and gooseberries at rates higher than one and a half cwt per acre.

Sulphate of Potash (48–50 per cent K_2O). Sulphate of potash is manufactured from the chloride and is therefore more expensive. Sulphate of potash, unlike muriate, shows little tendency to pick up moisture from the atmosphere. This makes it easier to store and to handle and spread.

Up to 100 units per acre can be applied as muriate during the winter but for spring applications sulphate of potash should be used for soft fruit (raspberries, red currants, gooseberries and, to a lesser extent, strawberries) which are damaged by chlorides and care should be taken that compound fertilizers should contain sulphate and not muriate of potash.

Sulphate of Potash-Magnesia (28 per cent K_2O, 5–8 per cent Mg). This fertilizer contains both sulphate of potash and magnesium sulphate and is useful where both potassium and magnesium are required. It has been used on the continent for some time under the name of 'Patentkali'.

There are also other fertilizers containing both potassium and magnesium in a range of concentrations (up to 40 per cent K_2O and 8 per cent Mg). Here the potassium is present as the chloride and the magnesium as sulphate. Potash Nitrate has already been mentioned under inorganic nitrogenous fertilizers.

MAGNESIUM-CONTAINING FERTILIZERS

The following are the principal sources of magnesium for fruit crops.

Agricultural Epsom Salts ($MgSO_4.7H_2O$, 10 per cent Mg). This is a coarsely crystalline material which tends to cake on storage. It is used mainly for foliar spraying but can also be used for soil application though it is an expensive type for the latter method.

Kieserite ($MgSO_4.H_2O$, 16 per cent Mg). Kieserite can be regarded as being a more concentrated form of Epsom salts as it contains less water of crystallization. Kieserite is insufficiently soluble to be used for foliar sprays but is very suitable for soil application being cheaper per unit of magnesium than Epsom salts.

Fertilizers containing both potassium and magnesium have been discussed in an earlier section.

All the above materials contain magnesium in the form of sulphate which is soluble in water although the degree of solubility varies somewhat with the form of crystallization.

Magnesian Limestone (analysis varies but good samples contain 9 per cent or more of Mg). In this material magnesium is present in a sparingly soluble and slowly available form (see later section Liming Materials).

Calcined Magnesite (MgO). This material is almost pure magnesium oxide containing 60 per cent Mg. In the granular form it is easily spread and, on a unit of Mg basis, is the cheapest source of magnesium though slower acting than Kieserite.

Bulky Organic Manures

There are a number of bulky organic manures used in fruit growing which are valued as much for the organic matter they provide as for the plant nutrients they contain. In some cases there is an overlap between materials in this section and those in the Organic Nitrogenous Fertilizers section, for example low grade shoddies may be regarded as bulky organic manures whereas high grade shoddies containing as much as 14 per cent nitrogen are a relatively concentrated source of organic nitrogen.

Farmyard Manure. Farmyard manure varies greatly in quality depending on:

(i) the kind of animal producing the manure;
(ii) the nature and the amount of litter used;
(iii) its state of decomposition;
(iv) storage conditions and length of storage.

Farmyard manure containing much undecomposed straw or other litter is known as 'long' manure whereas manure that has been allowed to break down in a heap is said to be 'short'. Long manure has beneficial opening effects on heavy soils or soils liable to cap on the surface but the nutrients are only released when the manure decomposes. When very strawy manure is ploughed in it is advisable to add one cwt of sulphate of ammonia or similar nitrogenous fertilizer per acre to supply the nitrogen needed for decomposition to take place. If this is not done, nitrogen will be taken from the soil to decompose the straw and this may lead to nitrogen deficiency in the crop. These considerations apply with even greater force when sawdust or wood shavings have been used as litter because such materials decompose very slowly in the soil unless the proportion of dung and urine to litter is high. Manures of this type containing much undecomposed sawdust or wood shavings are best stacked with the addition of nitrogenous fertilizer to aid decomposition before application to the soil.

Farmyard manure varies considerably in its content of manurial constituents but, on average, it contains about 0·5 per cent nitrogen (N), 0·25 per cent phosphoric acid (P_2O_5), 0·5 per cent potash (K_2O). Not all these nutrients are readily available to crops. Recent findings for agricultural crops suggest that in the first year after application 10 tons per acre of 'average' farmyard manure will provide the equivalent of about 1½ cwt sulphate of ammonia, 2 cwt superphosphate and 1¼ cwt of muriate of potash.

Deep Litter and Broiler House Poultry Manure. These materials are also extremely variable in composition depending on such factors as the source of litter and, more especially, on the time the manure has been in the houses. Sources of litter include straw, peat, sawdust and wood shavings. Difficulties that arise from the depletion of available nitrogen when sawdust or wood shavings are added directly to the soil can be overcome if these materials are first turned into well decomposed deep litter manure.

Well made samples can be regarded as being a bulky organic manure with much higher percentages of nitrogen, phosphate and potash than farmyard manure. Samples vary considerably in composition, for example from 1·0–2·4 per cent nitrogen (N), 0·4–3·4 per cent phosphoric acid (P_2O_5) and 0·8–1·9 per cent potash (K_2O).

Straw. Straw is sometimes used as a surface mulch to conserve moisture. When straw merely rests on the surface of the soil in this way, decomposition proceeds slowly and difficulty is not experienced with locking up of nitrogen. By comparison, when straw is worked into the soil, extra nitrogenous fertilizer must be given to avoid subsequent nitrogen deficiency in crops. One side-effect of applying heavy straw mulches is that, when they decompose, they provide substantial amounts of potash.

Sawdust. The main practical use of sawdust in fruit growing is as a mulching material, particularly for soft fruits and for newly planted fruit trees.

Shoddy. This is a wool waste from the mills. The highest grade obtainable from pure wool contains 12–16 per cent nitrogen but there is usually some admixture of cotton and dirt which gives rise to the most commonly available medium grade materials containing 5–10 per cent nitrogen and sometimes poor samples with as little as 2·5 per cent nitrogen. An analysis for nitrogen is clearly essential. The physical condition is also important, for a loose friable sample can be spread much more evenly than a matted one. Oily samples may break down slowly in the soil and release nitrogen slowly. Common dressings range from ½ a ton per acre for the best quality to 1–2 tons per acre of the more common grades. Two tons per acre of a medium grade shoddy will provide as much organic matter as 10 tons of farmyard manure and much more nitrogen. Materials composed largely of cotton waste contain very little nitrogen and can actually cause a temporary locking up of nitrogen if they are ploughed into the soil.

Spent Mushroom Compost. This material, which is readily obtainable in some areas, is of variable composition mainly depending upon the proportion of casing soil which it contains. The following figures show the range of composition which may be expected:

	N	P_2O_5	K_2O
		Units per wet ton	
Farmyard manure	11	4	11
Spent horse manure compost (excluding soil)	16	5	18
Spent horse manure compost (including soil)	8	3	9

The organic matter of spent mushroom compost is generally more decomposed than in strawy farmyard manure and hence it may be less valuable for improving heavy soils or for the prevention of capping on soils which are subject to this trouble. Another point is that where chalk and peat are used as casing material and the casing is subsequently mixed with the spent compost, the resultant product may contain excessive amounts of lime for fruit soils.

Apart from these provisos, spent mushroom compost can be a useful source of organic matter.

Sewage Sludge. Sludges from sewage disposal works are occasionally made available at low cost. These materials supply nitrogen and phosphorus but little potassium. They are not good sources of organic matter for soil improvement because of the nature of their organic content. Sewage sludges also often contain high concentrations of heavy metals such as copper, zinc and nickel and may also contain considerable quantities of boron, all of which are toxic to plants. For these reasons sewage sludge cannot be recommended for use on fruit.

Liming Materials

Under the Agriculture Act 1970, Part IV, suppliers of liming materials are required to provide a statement giving the Neutralizing Value of the lime and this provides a convenient method of comparing one liming material with another. The Neutralizing Value is the weight of pure burnt lime which is equivalent to 100 lb of the liming material under consideration. Thus a statement that a ground limestone has a Neutralizing Value (N.V.) of 53 means that 100 lb of the limestone is equivalent (so far as neutralizing soil acidity is concerned) to 53 lb of pure burnt lime (calcium oxide, CaO). Although there is quite a wide range of liming materials on the market, ground limestone, ground chalk and ground magnesian limestone are the forms most likely to be of interest to the fruit grower.

Details of other grades can be found in Bulletin No. 35: *Lime and Liming* (HMSO 36½p net).

Ground Limestone and Ground Chalk (Calcium carbonate, $CaCO_3$). These are limestone or chalk ground to a fine powder. A good sample is almost pure calcium carbonate and has a Neutralizing Value of about 55. Ground limestone and ground chalk can be stored for an indefinite period without fear of deterioration of the material or destruction of the bags. They are much less unpleasant to handle than quicklime and are very suitable for bulk handling and mechanical spreading.

The quality of ground limestone and chalk depends on the quality of the stone used and on the fineness of grinding. There is a limit to the fineness of grinding, beyond which the cost makes the price prohibitive. Furthermore, very fine limestone flour is difficult to distribute, either by hand or machine. The Statutory Statement given to the buyer shows, in addition to the Neutralizing Value, the percentage of the materials which will pass through a prescribed sieve (British Standard Test Sieve, mesh number 100). A product with 40 per cent passing this sieve is regarded as a good standard for the normal rates of dressing. A reasonably coarse product is cheaper to produce, passes more easily through a drill and is not so rapidly washed out of the soil.

It should be noted that ground limestones vary somewhat in colour but this is of no importance so far as their value as liming materials is concerned.

Ground Magnesian Limestone (calcium and magnesium carbonates, $CaCO_3$ and $MgCO_3$). The main deposit of magnesian limestone in this country is in a belt from Durham to Nottinghamshire although there are smaller deposits elsewhere, for example in the Forest of Dean. It is generally buff in colour and contains both calcium and magnesium carbonates but the proportion of magnesium carbonate varies from 4 to 45 per cent. Magnesian limestones containing a high percentage of magnesium carbonate are sometimes known as dolomite.

Magnesium carbonate is slightly more efficient in neutralizing soil acidity than calcium carbonate but, as all magnesian limestones contain a substantial percentage of calcium carbonate, they may to all intents and purposes be regarded as the equivalent of ordinary ground limestone and chalk. Magnesian limestone has special uses on acid soils which are also deficient in magnesium. For this purpose a magnesian limestone with a high magnesium carbonate content should be chosen, a minimum of 30 per cent magnesium carbonate (equivalent to approximately 9 per cent Mg) would be a reasonable standard. Forty per cent passing through the 100 mesh sieve is also a satisfactory degree of fineness for ground magnesian limestone.

Compound Fertilizers

A wide range of compound fertilizers can be obtained including some specially prepared for fruit crops. Compounds are available which contain nitrogen, phosphates and potash (so-called 'complete' compounds), nitrogen and phosphate only, nitrogen and potash only and phosphate and potash only. Some have magnesium salts added, while in others part or all of the nitrogen and phosphate may be derived from organic sources. The usual materials in the readily available compound fertilizers are sulphate of ammonia, ammonium phosphates, ammonium nitrate, ordinary or triple superphosphate and sulphate or muriate of potash.

Most compound fertilizers are sold in granular form and this makes for ease of storage, handling and distribution. Moreover, the use of compounds enables the grower to give two or more plant nutrients at a single application. Disadvantages of using compounds stem from the fact that they tend to be a little more expensive than the corresponding 'straight' fertilizers.

Compound fertilizers vary considerably in concentration, for example a compound containing 10 per cent N, 10 per cent P_2O_5 and 15 per cent K_2O has 35 plant food units per cwt, whereas a compound containing 13 per cent N, 13 per cent P_2O_5 and 21 per cent K_2O has 47 units per cwt (see section below on Evaluation of Fertilizers for an explanation of plant food units.) The more concentrated compound fertilizers save labour in handling and distribution.

Trace Elements

The principal trace elements to be applied to fruit crops are iron and manganese; they should only be applied when a deficiency has been confirmed. The treatment of these deficiencies is discussed in the section 'Mineral Requirements of Fruit' (see pages 21–31).

Evaluation of Fertilizers

It is sometimes necessary for the fruit grower to carry out simple fertilizer calculations, for example if one 'straight' fertilizer is to be substituted for another or to decide whether a compound fertilizer is equivalent to certain 'straights', also in connection with questions of balance of nutrients in fertilizers. Such calculations are best carried out using the plant food unit, that is one per cent of one cwt of the particular plant nutrient concerned. Thus sulphate of ammonia (21 per cent N) contains 21 units of N per cwt of fertilizer; one of the ammonium nitrate/lime fertilizers (the 25 per cent N grade) contains 25 units of N per cwt of fertilizer. Therefore, 5 cwt of sulphate of ammonia (105 units of N) are almost equivalent to 4 cwt of the ammonium nitrate/lime fertilizer (100 units of N) so far as nitrogen content is concerned.

Similar considerations apply to other fertilizers.

Superphosphate (19 per cent water-soluble P_2O_5) contains 19 units of water-soluble P_2O_5 per cwt.

Muriate of potash (60 per cent K_2O) contains 60 units of K_2O per cwt.

Sulphate of potash (50 per cent K_2O) contains 50 units of K_2O per cwt.

This system can be used to calculate the equivalent of compound and 'straight' fertilizers. Say a fruit grower wishes to apply a compound fertilizer equivalent to 2 cwt sulphate of ammonia, 2 cwt superphosphate and 1 cwt muriate of potash. The 'straight' fertilizers will provide the following:

2 cwt sulphate of ammonia (21 per cent N)—42 units of N;
2 cwt superphosphate (19 per cent P_2O_5)—38 units of P_2O_5;
1 cwt muriate of potash (60 per cent K_2O)—60 units of K_2O.

It will be noted that the ratio of N to P_2O_5 to K_2O (42:38:60) approximates to 1:1:1½ and this is known as the Plant Food (PF) Ratio. PF ratios for compound fertilizers are usually stated by the manufacturer so that it is only necessary in this case to obtain a compound fertilizer of PF ratio 1:1:1½. There are a large number of such fertilizers available, for example, one containing 12 per cent N, 12 per cent water-soluble P_2O_5 and 18 per cent K_2O and another containing 10 per cent N, 10 per cent water-soluble P_2O_5 and 15 per cent K_2O.

3½ cwt of the 12:12:18 fertilizer will provide 42 units of N, 42 units of P_2O_5 and 63 units of K_2O.

4 cwt of the 10:10:15 fertilizer will provide 40 units of N, 40 units of P_2O_5 and 60 units of K_2O.

Thus 3½ cwt of the first compound fertilizer and 4 cwt of the second are equivalent for all practical purposes to the 'straight' fertilizers.

So far as 'straight' fertilizers are concerned, the following are convenient equivalents:

5 cwt sulphate of ammonia (21 per cent N) contain the same amount of nitrogen as 4 cwt of ammonium nitrate/lime (the 25 per cent N grade) or 3 cwt ammonium nitrate (34 per cent N) or 6½ cwt of nitrate of soda (16 per cent N).

5 cwt superphosphate (19 per cent P_2O_5) contain the same amounts of phosphoric acid as 2 cwt of triple superphosphate (47 per cent P_2O_5).

1¾ cwt muriate of potash (60 per cent K_2O) contain the same amount of potash as 2 cwt sulphate of potash (50 per cent K_2O).

In addition to the above type of calculation it is also very desirable to be able to compare the relative prices of the same plant nutrient in different 'straight' fertilizers and also to be able to compare the costs of different compound fertilizers or fertilizer mixtures. These comparisons are made by obtaining the price per plant food unit of each nutrient concerned—nitrogen (N), phosphoric acid (P_2O_5) or potash (K_2O). The unit price of a plant nutrient is obtained by dividing the cost per cwt of the fertilizer by the percentage content of the plant nutrient in the fertilizer.

For example, a sample of sulphate of potash containing 50 per cent of potash (K_2O) costs £1·70 per cwt delivered. The unit price of potash is therefore $\frac{\text{cost per cwt}}{\text{percentage content of } K_2O}$ that is £1·70 divided by 50 or 3·4p per unit. In working out the unit price, the cost of the fertilizer should include cost of delivery. One obvious reason for including these charges is that low-grade fertilizers and bulky manures may, in the long run, prove quite expensive in supplying plant nutrients when heavy haulage and spreading costs are involved.

When the comparative cost of a compound fertilizer is being considered, use can be made of the unit price of the nutrients as obtained from the cost of the equivalent 'straight' fertilizers. When the unit price of each nutrient has been worked out the comparative cost of the compound can be obtained as in the following example:

A compound fertilizer contains 14 per cent nitrogen (N), 14 per cent water-soluble phosphoric acid (P_2O_5) and 21 per cent potash (K_2O). It is desired to compare its cost with that of equivalent 'straight' fertilizers. The equivalent 'straight' fertilizers are assumed to be sulphate of ammonia (for N), superphosphate (for P_2O_5) and muriate of potash (for K_2O).

No. and Unit Price	*Value per cwt* £
14 units of nitrogen at 6p per unit	0·84
14 units of water-soluble phosphoric acid at 9p per unit	1·26
21 units of potash at 3·2p per unit	0·67
	£2·77

To the figure thus obtained (that is £2·77 in the example given) a reasonable amount should be added to cover the manufacturer's cost of mixing, granulating (in the case of granulated fertilizers), bagging and other overhead expenses.

Care must be taken in applying the unit price system if the compound fertilizer contains special constituents of value, for example if a compound fertilizer for soft fruit crops contains sulphate of potash then obviously the unit price of K_2O in sulphate of potash must be used for purposes of calculation. Similarly when organic forms of nitrogen are required, then nitrogen in this form in a compound fertilizer must be valued at a higher unit price than nitrogen in sulphate of ammonia.

It should also be remembered that potash nitrate, sulphate of potash-magnesia, meat-and-bonemeal and so on contain more than one plant nutrient; in these the unit price of each nutrient has to be taken into account in assessing the relative value of different compounds. A basis of comparison on the lines of that suggested for compound fertilizers would be suitable.

Manuring of Different Fruit Crops

THE nitrogen requirements of fruit crops grown in Britain differ considerably with the type and age of the crop, the soil type, system of management and climate. Potassium and magnesium requirements, however, are similar for all crops in that the optimum soil levels are similar. But in spite of this the severity of the effects of deficiency varies between crops. Also different soils need different amounts of fertilizer to attain optimum levels.

The majority of fruit crops have shown little if any response to phosphate fertilizers in experiments but soft fruits tend to be a little more responsive than tree fruits. In spite of this lack of response it is sensible to maintain a satisfactory level of phosphate in the soil as even a small response would be economically worthwhile. Also where grass swards are grown in orchards phosphate may be needed for their satisfactory establishment and growth.

On soils seriously deficient in phosphate, potassium or magnesium it is necessary to apply fertilizers containing the deficient nutrient or nutrients every year. However on soils only slightly deficient they can be applied every two years and maintenance dressings need only be applied once every three years.

Fertilizer requirements will be influenced by whether applications of farmyard or other bulky organic manures are used. Mulches of these or of other materials such as straw and also irrigation can profoundly affect growth in response to fertilizers.

The habit of growth and cropping of the plant will also have an important influence on its nutrient requirements. Top fruits when first planted must form a framework of branches on which the subsequent crops are borne and once cropping has started it is necessary to maintain a balance between growth and cropping. The black currant fruits largely on new wood so that it is essential to maintain the vegetative vigour of the bushes if yields are to be satisfactory. This applies also to raspberries and loganberries, both of which fruit on canes formed in the previous season. Gooseberries and red currants fruit mainly on spurs on older wood and so there is not the same need for new annual growth of maiden shoots or for the heavier applications of nitrogenous fertilizers given to black currants.

While there is much information available from experiments in this country on the manuring of apples and most of the soft fruits, little has been done experimentally to determine the fertilizer requirement of pears and the stone fruits, in particular of cherries. The recommendations for the use of fertilizers for these crops are, therefore, to some extent based on experience in the orchard.

Rates of phosphate, potash and magnesium applied should be related to soil analysis and recommendations are summarized in the following tables. When severe deficiences are encountered before establishing new orchards it is important to plough down generous applications of fertilizers since applications to the surface move into the root zone of crops only very slowly.

APPLES

A balanced nitrogen supply is most important for the satisfactory maintenance of growth and cropping of apple trees, and a deficiency of this element not only restricts growth but encourages any tendency towards biennial bearing. Adequate nitrogen is important for the satisfactory development of fruit buds and has been shown to have an important effect on the quality and the set of the flowers. It is also known to influence the rate of pollen tube growth.

Recommended rates of nutrients per annum in units per acre

	P				K					Mg			
Nutrient Index	0	1	2	Over 2	0	1	2	3	Over 3	0	1	2	Over 2
Crop	Units P_2O_5				Units K_2O					Units Mg			
Apples (dessert) Apples (culinary) Cherries Pears Plums	60	30	15	Nil	200	150	75	25	Nil	50	35	25	Nil
Red currants Black currants Gooseberries Raspberries Loganberries Blackberries Strawberries	90	60	30	Nil	200	150	100	50	Nil	50	35	25	Nil

Notes: (1) IT IS ADVISABLE TO TAKE SAMPLES FROM 6–12 in. DEPTH AS WELL AS FROM 0–6 in. WHEN SAMPLING TOP FRUIT SOILS.

(2) When a K index of 0 is encountered before planting, a heavy extra dressing of potash (up to 300 units K_2O per acre) should be ploughed down, in addition to the normal application recommended.

(3) At P indices greater than 0, K indices greater than one and at all Mg indices, twice the annual rates can be applied every other year.

(4) For top fruit where the K index for the 0–6 in. sample is less than 2 and that of the 6–12 in. sample is 0 the potash recommendations should be increased by 50 units K_2O.

(5) Muriate of potash (60% K_2O) is suitable for tree fruits except in coastal areas where soil chlorides are likely to be high and sulphate of potash should be used.

(6) Muriate of potash (60% K_2O) is suitable for black currants and at all recommended rates up to 100 units K_2O per acre applied in winter for other soft fruit. Red currants, raspberries and gooseberries are more likely to be damaged by chlorides and care should be taken to see that excessive amounts are not applied. In coastal areas where soil chlorides are likely to be high, sulphate of potash should be used for all crops.

(7) *Mulches.* Where FYM is used the appropriate reductions in the rates of nutrients applied should be made. (See page 43). Where straw is used, the rates of potash applied should be reduced by 15 units for every ton of straw.

Under arable conditions the breakdown of organic matter in the soil may provide most of the nitrogen requirement so that when trees are first planted and maintained under clean cultivation, with or without mulches, there is a need for only small applications of nitrogenous fertilizers. When an orchard is grassed down, the sward competes severely with the trees for nitrogen supplies and it is essential to apply sufficient nitrogen to allow for this. At the time of grassing down it is also advisable to maintain a mulch around the trees. The maintenance, with the use of herbicides, of a grass free area round the tree also helps to reduce competition with the tree for both water and nutrients.

When using nitrogenous fertilizers enough should be applied to maintain good leaf size and colour and satisfactory new growth. Where trees are excessively vigorous and failing to crop, applications of nitrogen may be temporarily withheld or reduced in amount. Similarly where irrigation is used or mulches are employed, the need for nitrogen is generally reduced.

As excessive nitrogen can adversely affect the colour of dessert apples, especially of varieties such as Worcester Pearmain, in which good colour is of first importance, very heavy applications of nitrogen should be avoided. Excessive nitrogen can affect fruit quality in other ways. If it leads to excessive vigour there will be a tendency for the production of fruit with a low calcium level with, in consequence, an increased risk of bitter pit and breakdown in storage. Highly nitrogenous fruit is also more susceptible to fungal rots. Although in general, cooking apples should be given larger nitrogen dressings than dessert apples excessive amounts can adversely affect storage quality.

The application rates of nitrogen will depend not only on the condition of the trees, the presence or absence of a grass sward with or without herbicide strip, but also on the soil and local climatic conditions. Lower rates may be used in areas of moderate to high rainfall than in the drier areas of the south-east and East Anglia. Lower rates can also be used on the heavier, more retentive soils than on lighter drier soils. The rates given in the following table can be used as a guide for average conditions:

Recommended Rates of Nitrogen

		Cultivated	Grass herbicide strip	Grass
		Units per acre per year		
Dessert	Summer rainfall less than 14 in.	40	60	100
	Summer rainfall greater than 14 in.	30	40	70
Culinary	Summer rainfall less than 14 in.	80	100	140
	Summer rainfall greater than 14 in.	50	70	100

When grassing down a plantation the initial dressings of nitrogen may be increased. Symptoms of nitrogen deficiency are often seen in the year following grassing down and in anticipation of this it is as well to step up the rate of nitrogenous fertilizers applied in the year in which the sward is established, and also to repeat the higher rate the following year. The likelihood of the trees suffering from nitrogen deficiency however is greatly reduced by keeping the soil surrounding them free of grass and weeds by means of herbicides. Foliar sprays of urea may be used to supplement the nitrogen supply to the trees at the time of grassing down when the uptake of nitrogen may be limited. The normal method of use is to add urea during the source of routine spraying at a rate equivalent to 5 lb per 100 gal (high volume). Where urea is used on its own, the addition of a wetter is necessary. To obtain a satisfactory response it is usually necessary to apply two to three sprays during the course of the growing season. Urea sprays throughout the growing season have also been found to be of value for trees carrying heavy crops, by helping to increase yields the following year. Urea sprays are of no value to trees which are not nitrogen deficient and may even impair fruit quality.

The requirements of apples for phosphorus are quite low and phosphorus deficiency has not been observed in apple orchards in Great Britain. Annual applications of phosphates are not usually necessary but if there is any real deficiency it should be remedied prior to planting particularly as it is now known that fruit low in phosphorus may break down in store. Where the orchard is to be grassed down an adequate level of phosphate is necessary to ensure satisfactory establishment and maintenance of the sward.

On soils inadequately supplied with potash young trees often show acute signs of potassium deficiency and unless this is remedied at an early stage their growth may be severely checked and this may persist for some years in spite of subsequent treatment. For this reason, and since potassium moves relatively slowly through the soil, it is especially important to remedy any soil deficiency prior to planting by ploughing in a suitable potassic fertilizer.

The need for potash applications usually continues throughout the life of the tree as substantial amounts of this element are removed in the crops. Because excessive applications of potassium may reduce the uptake of magnesium by the tree, care should be taken to maintain a satisfactory balance between these two elements. Also excess potassium tends to increase the risk of the occurrence of bitter pit.

A regularly mown grass sward or a mulch round the trees tends to increase the amount of available potassium in the soil due to potassium being brought up from the subsoil by grass roots or from potassium in the mulch. Once a satisfactory level of potassium in the soil has been reached annual applications of 25–75 units per acre of muriate or sulphate of potash are ample and higher dressings are wasteful. When soil levels are high applications of potash are not needed every year and instead may be given in alternate years or after longer periods.

Apples are sensitive to magnesium deficiency and the relative proportions of potassium and magnesium need to be considered when applying fertilizers. Where potassic fertilizers are used excessively under soil conditions of low magnesium, deficiency symptoms of the element may quickly appear. Young actively growing trees often show such symptoms and this deficiency is also often seen in older heavy cropping trees.

If soil analysis shows a deficiency of magnesium, routine soil dressings of kieserite or other magnesium containing materials should be applied until the deficiency has been corrected. When soil levels of potassium are very high applications of potash fertilizers should be withheld until a satisfactory balance between magnesium and potassium has been obtained. Leaf symptoms of magnesium deficiency can be rapidly controlled by three or more foliar sprays of 2 per cent magnesium sulphate (Epsom salts) but spraying should be followed by correction of any soil deficiency.

Pears

In general pears respond to a higher level of nitrogen than dessert apples and can be given more liberal treatment without deterioration in the quality of the fruit. Many successful bush pear plantations have been kept under cultivation and manured freely especially with bulky organic manures, but there is an increasing tendency to use grass swards in pear orchards with a herbicide strip in the row to minimize competition from the sward. When pears are grown under overall grass, adequate dressings of nitrogen must be given and the grass kept short by regular mowing as otherwise the growth of the trees may be severely checked and yields and size of individual fruits adversely affected. Pear trees which have been badly checked in growth usually take longer to recover then apples. The risk of a severe check is greatly reduced by the maintenance of a weed free strip of ground round the trees by the use of herbicides.

Mulching helps to maintain satisfactory growth and cropping of pears when grown under either arable conditions or grass, and may consist of straw or dressings of dung or other bulky manures. A suitable rate is 10–15 tons per acre farmyard manure applied as a mulch around the trees.

The rates of nitrogen below are given as a guide for average conditions, but should be reduced if dung or other manure is applied.

Recommended Rates of Nitrogen

	Cultivated	Grass herbicide strip	Grass
	Units per acre per year		
Summer rainfall less than 14 in.	100	120	160
Summer rainfall greater than 14 in. . . .	70	80	110

Pears are less susceptible than apples to potassium or magnesium deficiency.

Plums

Plums have usually been grown most successfully under arable conditions of high nitrogen and plentiful soil moisture during the cropping period. When grown in grass they have often shown varied degrees of nitrogen deficiency even when liberal dressings of nitrogenous manures have been given. Competition for water when the fruit is swelling can be severe under a grass sward. Though this is less where the grass is kept short by mowing, the size of the fruit may suffer in periods of drought.

In addition to needing a high level of nitrogen, plums usually benefit considerably from dressings of farmyard or other bulky manures, though the cost of farmyard manure is rarely justified under present conditions except for mulching young trees. Recommended rates of nitrogen are similar to those for pears given in the previous section.

In general most varieties of plums are less susceptible to potassium deficiency than apples though there are one or two notable exceptions, in particular Purple Pershore, which is highly susceptible. Varieties subject to deficiency often show a forward rolling of the leaf margins.

Plums are not very susceptible to magnesium deficiency.

Cherries

The best crops of sweet cherries are usually grown on deep fertile loams, grass swards being allowed to develop once the trees have become established. Cherries are more responsive to nitrogen than dessert apples but less so than pears or plums. The following rates of nitrogen are given as a guide for average conditions:

Recommended Rates of Nitrogen

	Cultivated	Grass herbicide strip	Grass
	Units per acre per year		
Summer rainfall less than 14 in.	70	90	120
Summer rainfall greater than 14 in. . . .	40	60	80

There is no evidence that organic manures are preferable to inorganic fertilizers for supplying the nitrogen requirements of cherries, though it is generally best to avoid excessive dressings of quick acting nitrogenous fertilizers. Mulches of farmyard or other bulky manures are valuable on dry soils, particularly in helping to maintain adequate soil moisture during the cropping period.

Cherries are not very susceptible to potassium deficiency.

Magnesium deficiency in nursery trees and especially in cherry root-stocks is of frequent occurrence though it is not usually of major importance in commercial orchards.

Black Currants

To produce heavy crops early in the life of a plantation and to continue to crop satisfactorily, black currant bushes must be able to grow sufficient new wood each year; for this reason they are notable for their response to farm-yard and other bulky organic manures, also to mulches of straw or compost which help to retain moisture in the soil. Black currants also respond well to irrigation particularly on soils of a light texture and under conditions of limited summer rainfall. Maximum yields cannot be obtained from the use of fertilizers alone. In fact there may be lack of response to increasing nitrogenous fertilizer dressings where water is a limiting factor. The best results are generally obtained by using mulches of farmyard manure or other organic material along the rows and supplementing these with ferti-lizers. As a general guide 120 units of nitrogen per acre are recommended in areas with a summer rainfall of less than 14 in. and 60 units per acre in wetter areas. Care must be taken to avoid applying fertilizers in excess to an area alongside the applied bushes. Often serious damage results from an overdose of fertilizer applied at the base of the bush.

Though phosphate deficiency has not been seen in the field in Britain, favourable responses to applications have been shown in some experiments. Any deficiency should be put right before planting. Often, where the soil under the bushes is mulched they are able to take up more phosphorus than where no mulch is used.

Black currants are moderately responsive to potash and the muriate is as suitable as the sulphate.

Magnesium deficiency symptoms are frequently observed in black currants and are accentuated where high rates of potash fertilizers have been used, but deleterious effects on vigour and cropping are usually small. Black currants are susceptible to soil acidity and the pH should not be allowed to fall below 6·0 and preferably should be maintained near to 6·5.

Red Currants

Red currant bushes fruit mainly on spurs located on the older wood and so there is not the same need, as with the black currant, to maintain a high state of vegetative vigour. For this reason once a framework of branches has been formed the requirement of nitrogen is relatively low. Mulches, however, will benefit young bushes and mature ones grown on drought susceptible soils. As a general guide 80 units of nitrogen per acre are recommended in areas where the summer rainfall is less than 14 in. and 40 units per acre in wetter areas.

An outstanding feature in the cultivation of red currants is their susceptibility to potassium deficiency. When acutely deficient in potassium, however, their reaction to potash fertilizers is relatively slow. As red currants are liable to chloride injury, muriate of potash and other potash salts containing chlorides should not be used in excess of 100 units per acre. If higher rates are needed the sulphate should be used.

Gooseberries

Cropping bushes of gooseberries should be maintained in a sufficiently vigorous state to ensure good crops, but over stimulation of shoot growth resulting in overcrowded bushes needing much pruning is to be avoided. Very vigorous shoots are also liable to be 'blown out' by wind. Gooseberries respond well to dressings of farmyard manure which by helping to conserve soil moisture ensure berries of good size and substance. This is of particular importance with dessert varieties such as Leveller where good berry size is essential.

The nitrogen fertilizer requirements of gooseberries are similar to those of red currants. They also resemble red currants in their susceptibility to both potassium deficiency and chloride injury.

Raspberries

The raspberry crop is borne on canes produced in the previous year so that high yields depend on the production each year of a satisfactory number of canes of reasonable vigour. For this reason any factor limiting growth such as a shortage of water during the summer may have an adverse effect on the crop for two successive seasons. Mulches of farmyard manure or straw are very beneficial both for aiding moisture retention by the addition of organic matter and the nutrients provided.

Nitrogen promotes vigorous growth and has an influence on the numbers of canes produced as well as on their size. Nitrogen in excess, however, can have undesirable effects such as the production of over-vigorous and lush canes which ripen poorly and may be killed back by frosts. Excessively tall canes with long internodes need to be cut back to a reasonable height to aid picking and so the weight of cropping is reduced. As a general guide 80 units of nitrogen per acre are recommended for young canes in areas with a summer rainfall of less than 14 in. and 40 units per acre in wetter areas. Once the canes are established the amount of nitrogen given should be reduced by 20 units.

Raspberries are susceptible to potassium deficiency, also to chloride injury and so muriate of potash should not be applied in excess of 100 units per acre per annum. If higher rates are needed the sulphate should be used.

Raspberries are very susceptible to iron and manganese deficiencies and so it is particularly important to avoid inducing these deficiencies by applying lime to the soil in such quantity as to increase the pH to 6 or more.

Blackberries and Loganberries

To crop regularly and heavily blackberries and loganberries need to produce each year an ample supply of well-ripened canes. To do this the plants should root freely and grow steadily throughout the season. There has been little research in this country into the manurial requirements of these crops but in many respects their nutrient requirements are similar to those of raspberries.

Strawberries

The strawberry may be grown and cropped for one year only or cropping may be extended for up to three years or more. Manuring will, therefore, vary according to the time the plantation is retained.

Where a dressing of 20–30 tons per acre of farmyard manure is ploughed into the land before planting, and provided any soil deficiency of potassium or phosphorus is remedied prior to planting, further fertilizer applications will not usually be needed where cropping is limited to the maiden year. In fact, in several experiments a pre-planting dressing of farmyard manure has provided all the needs of the crop over several years.

While nitrogen is important for strawberries recent experiments have not shown a response in cropping to nitrogenous fertilizers that might have been expected, though in some instances lack of nitrogen reduced the vigour of the plants in their third year of cropping. The plants used in these experiments were from healthy virus tested clones which normally make vigorous growth in their early years and so may not need added nitrogen. Where growth is weak due to virus infection or is a varietal characteristic there will be a greater need for nitrogenous fertilizers. It is generally recommended that no nitrogen should be applied in high rainfall areas and none until the third year where the summer rainfall is less than 14 in.

Excess nitrogen can be detrimental to strawberries, resulting in excessive vigour and even a reduction in cropping due to the vegetative state limiting flower initiation. Fruit from vigorous plants is more susceptible to grey-mould (*Botrytis*), and heavy applications of nitrogen encourage weed growth.

Strawberry growers have shown some preference for organic forms of nitrogen such as hoof-and-horn or meat-and-bone manure owing to their alleged safety on account of slower release of nitrogen for the plants, but inorganic forms when used properly have given equal results.

The importance of potassium for strawberries has been shown in many experiments and in the absence of dung its omission has generally had disastrous effects on vigour, longevity and yield. Annual applications of potash fertilizer are therefore needed on deficient soils especially when adequate dressings of farmyard manure are not available. Sulphate of potash is generally preferred to forms containing chloride when heavy dressings are required.

The magnesium requirement of strawberries is not large but this deficiency sometimes occurs on poor soils such as those overlying gravels and where excessive applications of potash fertilizers have been used.

Strawberries are tolerant of fairly acid soil conditions but sometimes fail on a calcareous soil where deficiencies of iron or occasionally manganese may be induced, the latter especially on soils with a high organic matter content. Liming above pH 6·0 should be avoided.

When applying fertilizers to strawberries the materials should be applied between the rows of plants and undue amounts should not be allowed to fall on the foliage.

Strawberries grown under glass or cloches, and those grown only to produce runners will need different fertilizer treatment.

Index

Printed in Scotland by Her Majesty's Stationery Office at HMSO Press, Edinburgh
Dd 506731 K16 3/75 (11814)